Introduction to Vortex Based Mathematics

Martha Pieper

Author's note:

This book is based on the mathematical principles discovered by Marko Rodin and focuses only on the basic principles of Vortex Based Mathematics. My aim with this book is to explain these mathematical principles in a simple and easy to understand way. To expand your knowledge of this subject, I recommend reading the extensive material written by Marko Rodin.

Martha Pieper

Contents

7 The doubling and halving circuit
10 Dividing by two
11 The magnetic numbers 6 and 3
12 The number 9 is the control in Vortex Based Mathematics
14 The multiplication
15 Vortex Based Mathematics multiplication tables
21 Reciprocal numbers
26 Division
29 Dividing by 7
31 The division tables
32 Dividing by 3 and 6
32 The number 9
33 Addition and subtraction
34 The family of numbers 1 4 7
35 The family of numbers 2 5 8
36 The family of numbers 9 3 6
38 The family of numbers 1 4 7 and 2 5 8
41 The numbers 3 6 9
43 The doubling circuit never breaks
47 The power of the number 10
48 Quantum Numerology
49 The quantum numerology map
55 In Vortex Based Mathematics everything revolves around the number 9
58 The family numbers: 1 4 7, 2 5 8, +3+9+6, -3-9-6
61 Reference

Introduction to Vortex Based Mathematics

Philosophy is written in the great book which ever lies before our eyes – I mean the universe – but we cannot understand it if we do not first learn the language and grasp the symbols, in which it is written. This book is written in the mathematical language…without which one wanders in vain through a dark labyrinth.

Galileo Galilei

The doubling and halving circuit

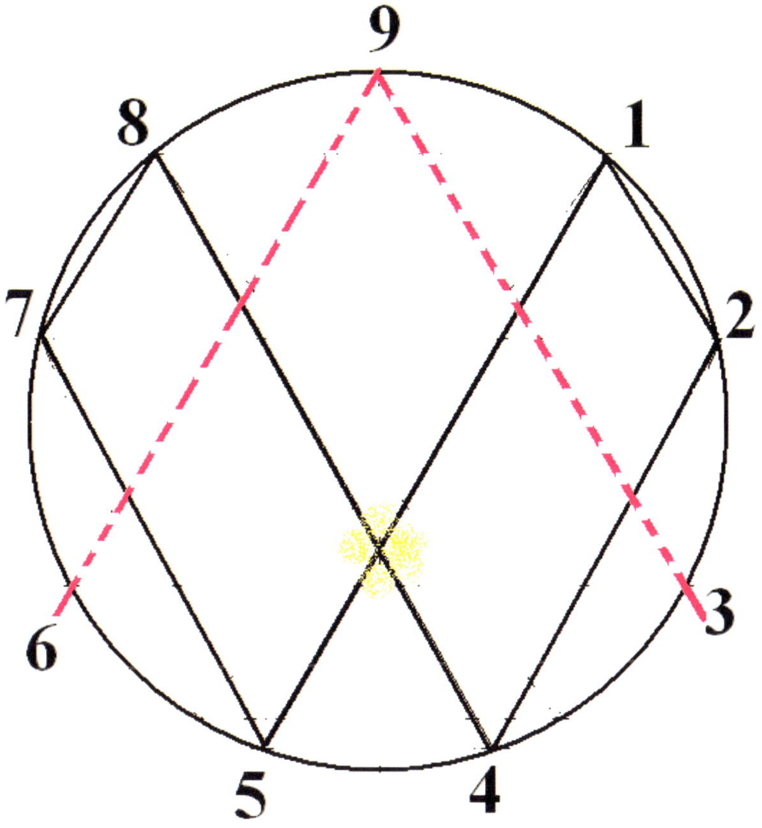

fig.1 The doubling and halving circuit, also call the symbol of enlighten or finger print of God by Marko Rodin

All the mathematical functions: addiction, subtraction, division, and multiplication can be solved simultaneously with the doubling and halving circuit. This symbol also models all the principles of physics, chemistry, genetics and biology. The numbers that exist in the world of creation are: 1,2,4,5,7 and 8, and they form the geometry in nature. The geometry in nature is essentially hexagonal. The numbers 3 and 6 are the magnetic numbers, and the number 9 is the emanation* coming from the center of the doubling circuit. The number zero does not exist in Vortex Based Mathematics. The number zero is defined as an absence or empty space.

The doubling and halving circuit

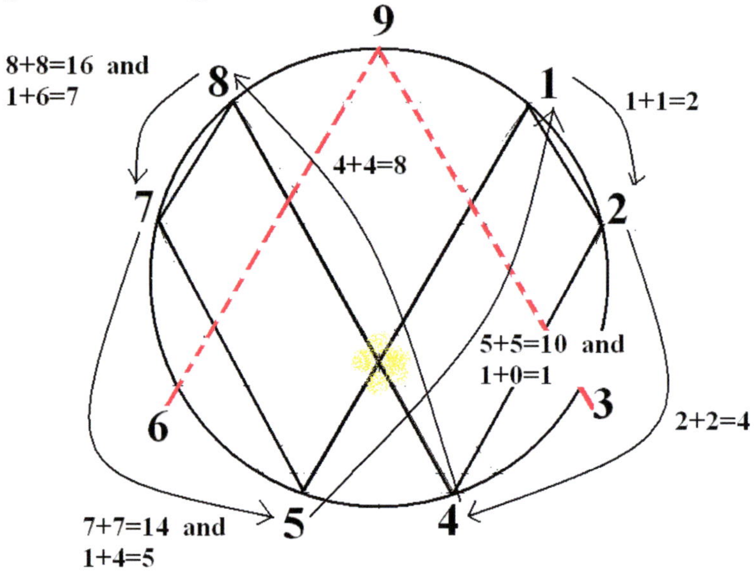

fig. 2

Starting with the number 1 and adding 1 is equal to 2, adding 2 to the number 2 is equal to 4, adding 4 to the number 4 is equal to 8, adding 8 to the number 8 is equal to 16 which is 1 plus 6 equal to 7, adding 7 to the number 7 is equal to 14 which is 1 plus 4 equal to 5, adding 5 to the number 5 is equal to 10 which is 1 plus 0 equal 1; here the cycle ends. We can continue infinitely with this process by doubling the numbers and adding the single digits, and it will always be equal to 1,2,4,8,7, and 5, see (fig. 2, 3 y 4).

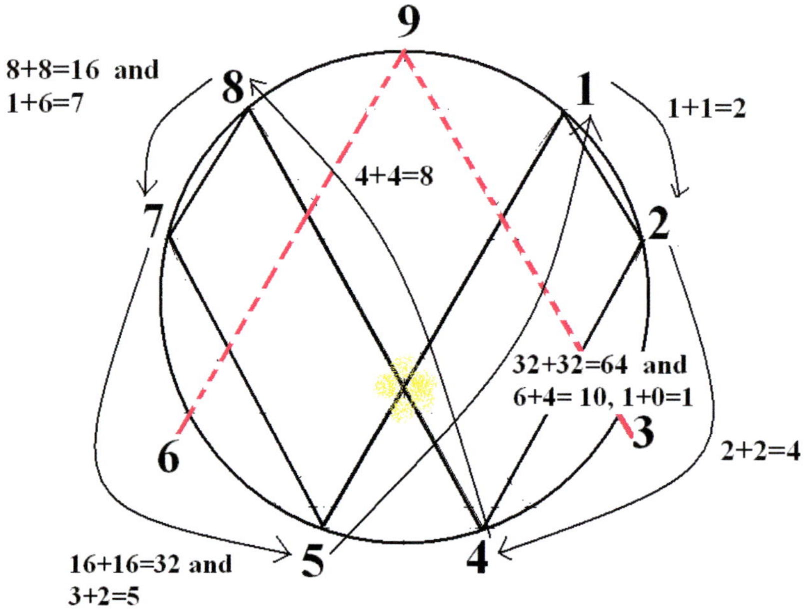

fig. 3

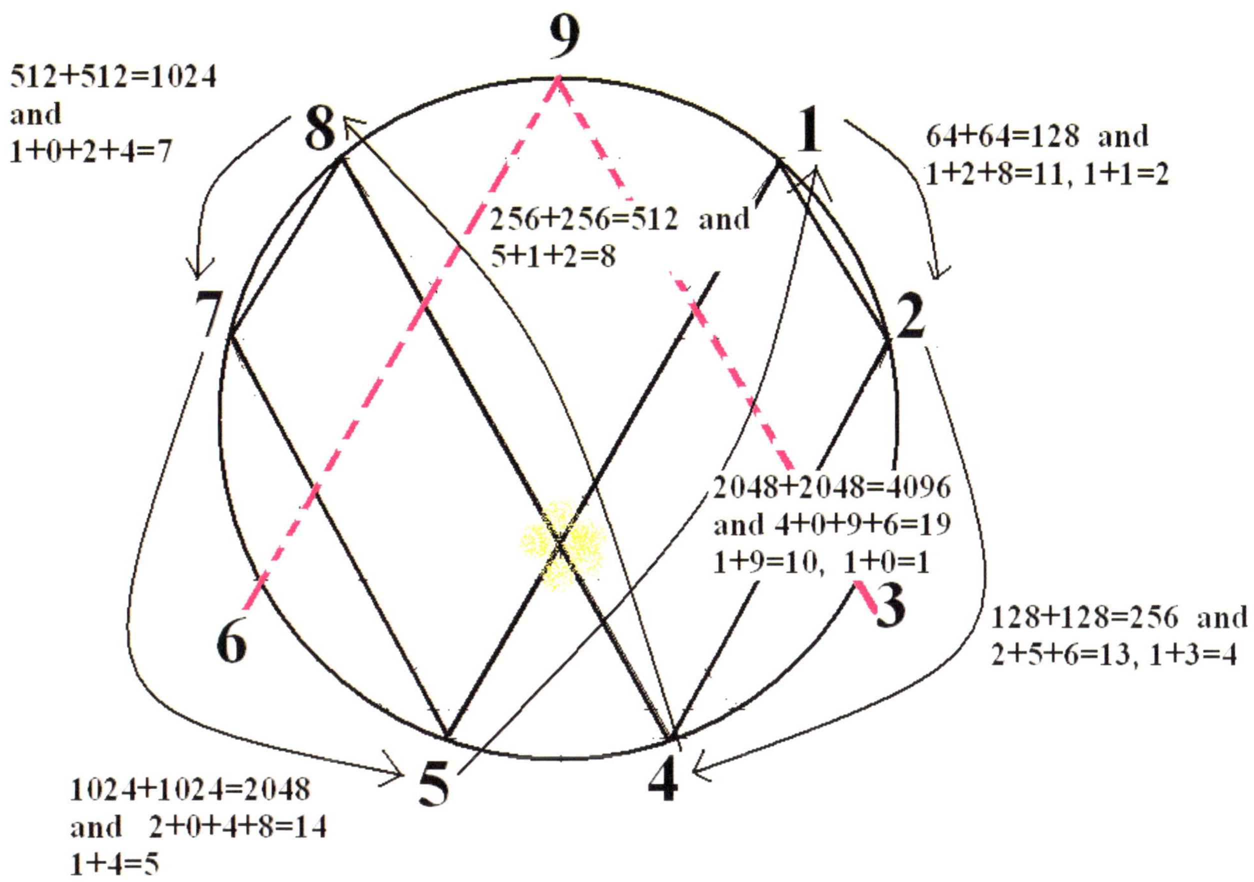

fig. 4

The doubling circuit shows how action, reaction and vibration occur in pairs. The cells division and the nuclear reaction, for example, occur in pair division. The number 64 is the culmination of a cycle, and after the number 64 the cycle will repeat infinitely, see (fig. 4). For example, in the embryo gestation, the division of cells begins in pairs until they reach the division 64. After the division number 64, the natural distinction of cells begins. At this point, it is where the cells identified with certain organ of the body.

Dividing by two

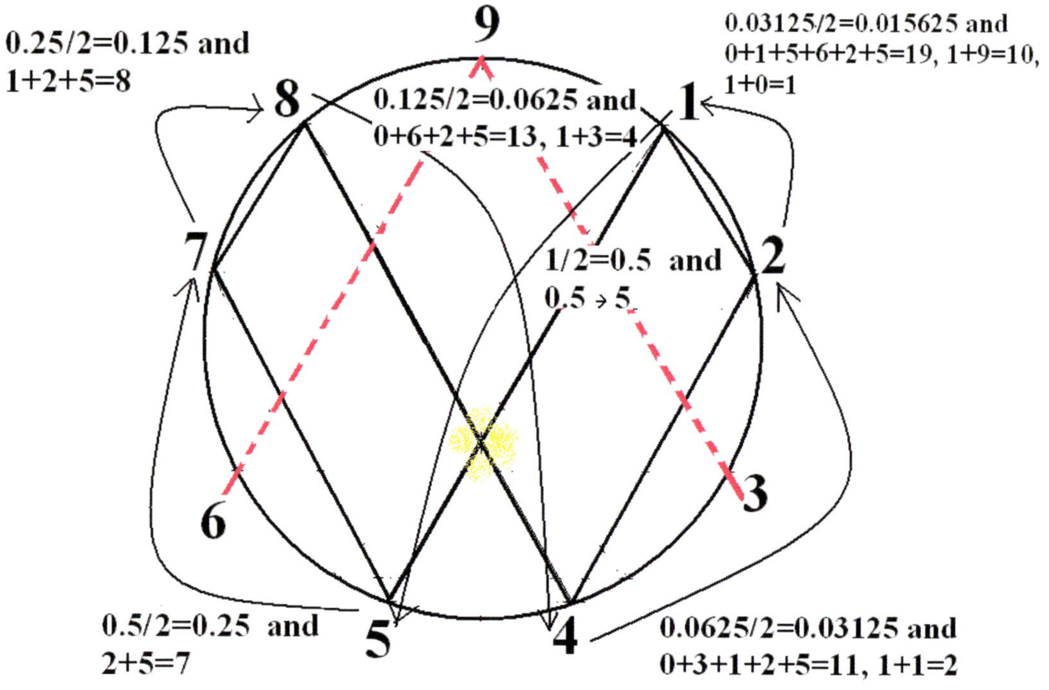

fig.5

In the opposite direction, 1 divide by 2 is equal to 0.5 equal to 5, 0.5 divided by 2 is equal to 0.25 and 2 plus 5 is equal to 7, 0.25 divided by 2 is equal to 0.125 and 1 plus 2 plus 5 is equal 8, 8 divided by 2 is equal to 4, 4 divided by 2 is equal to 2, 2 divided by 2 is equal to 1, and the cycle ends, see (fig. 5). We can continue dividing infinitely by two, and the sum of the digits of the decimal fractions always are going to add the digits: 5,7,8,4,2 and 1, see (fig. 6).

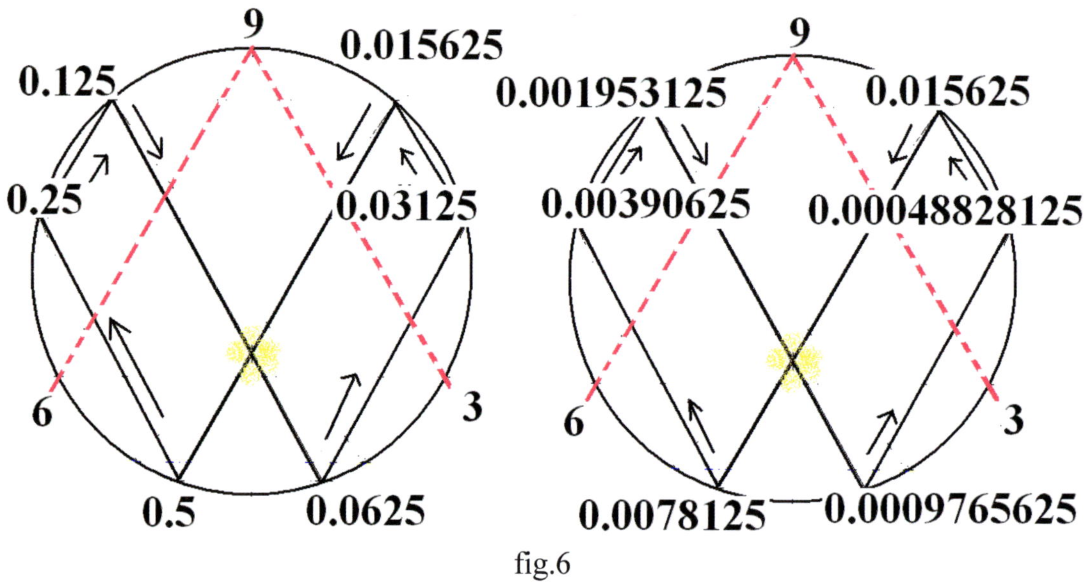

fig.6

The magnetic numbers 6 and 3

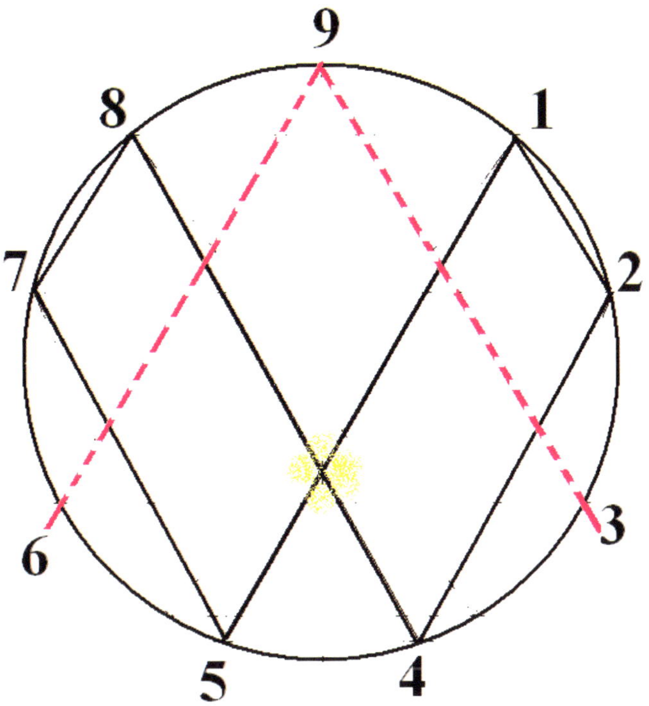

fig. 7

The numbers 6 and 3 oscillate when they are being doubling. For example, 3 plus 3 equals 6, and 6 plus 6 is equal 12 which is 1 plus 2 equal 3; and this happens infinitely. The numbers 6 and 3 oscillate, 6 to 3 and 3 to 6 successively and infinitely.

3+3=6	6+6=12 ~ 1+2=3
12+12=24 ~ 2+4=6	24+24=48 ~ 4+8=12 ~ 1+2=3
48+48=96 ~ 9+6=15	96+96=192 ~ 1+9+2=12 ~ 1+2=3
~ 1+5=6 …………..	…………………………………

When the numbers 6 and 3 are divided by two they also oscillate back and for; 3 divided by 2 is equal to 1.5 and 1 plus 5 is equal to 6, 6 divided by 2 equals 3 and this oscillation happens infinitely.

3/2=1.5 ~ 1+5=6	1.5/2=0.75 ~ 7+5=12 ~ 1+2=3
0.75/2=0.375 ~ 3+7+5=15	
1+5=6 …………………	0.375/2=0.1875 ~ 1+8+7+5=21
	2+1=3 ……………………….

The number 9 is the control in Vortex Based Mathematics

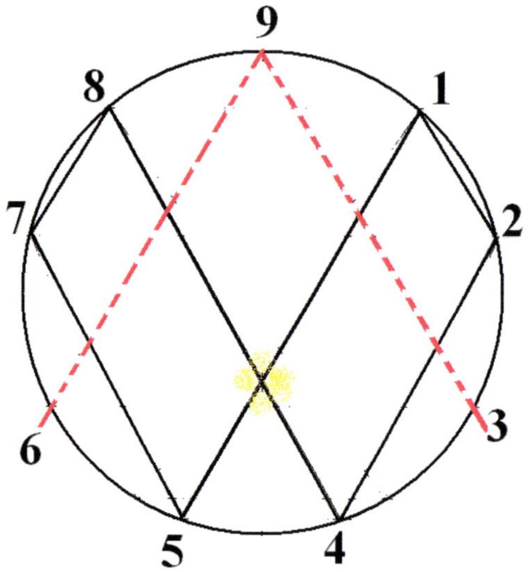

fig. 8

Doubling the number 9 by itself always add 9

9+9=18 ~ 1+8=9
18+18=36 ~ 3+6=9
36+36=72 ~ 7+2=9
72+72=144 ~ 1+4+4=9 …………continues infinitely.

Dividing the number 9 by two will always add 9

9/2=4.5 ~ 4+5=9
4.5/2=2.25 ~ 2+2+5=9
2.25/2=1.125 ~ 1+1+2+5=9
1.125/2=0.5625 ~ 5+6+2+5=18, 1+8=9…...continues infinitely.

The number 9 has no parity. If the number 9 is doubled or divided by two, it is always equal to itself. The number 9 never changes, because it is a linear emanation* coming from the center of the doubling circuit. The doubling circuit is actually a vortex.

*Note: The concepts of Vortex Based Mathematics will be understood as the reader reads more about this subject, this book is presenting only the basic concepts.

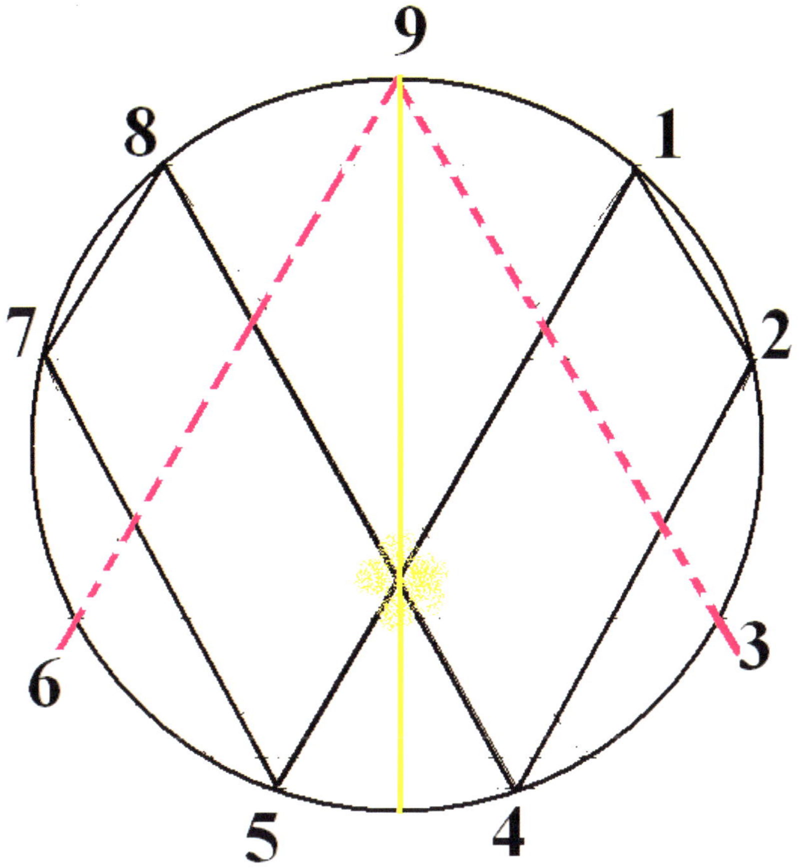

fig. 9

The number 9 is the linear emanation* coming from the center of the doubling circuit that separates and polarizes the numbers: 1,2,4,5,7 and 8. The number 9 transforms the doubling circuit numbers into images. Thus, the numbers 8 and 1, 7 and 2, 5 and 4 are images. We can also observe that these pairs of numbers: 8 and 1, 7 and 2, 6 and 3, 5 and 4, add up to 9.

8 +1 = 9
7 +2 = 9
6 + 3 = 9
5 + 4 = 9

*Note: The concepts of Vortex Based Mathematics will be understood as the reader reads more about this subject, this book is presenting only the basic concepts.

The multiplication

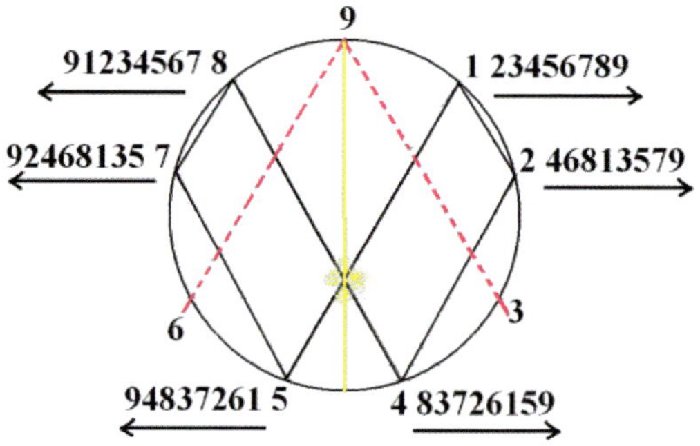

fig.10

Vortex Based Mathematics multiplications are reduced to a single digit, for example 2x8=16, and 1 plus 6 equal 7. As we mentioned earlier, the number 9 creates the invisible line that separate and polarize the numbers: 1,2,4,5,7 and 8 and converts them into images. The number 8 and 1 are images, because when they multiply by the digits: 1,2,3,4,5,6,7,8 and 9 behave like mirror images see, (fig. 10 and 11).

1x8=8	1x1=1
2x8=16~1+6=7	2x1=2
3x8=24~2+4=6	3x1=3
4x8=32~3+2=5	4x1=4
5x8=40~4+0=4	5x1=5
6x8=48~4+8=12~1+2=3	6x1=6
7x8=56~5+6=11~1+1=2	7x1=7
8x8=64~6+4=10~1+0=1	8x1=8
9x9=72~7+2=9	9x1=9

1x7=7	1x2=2
2x7=14~1+4=5	2x2=4
3x7=21~2+1=3	3x2=6
4x7=28~2+8=10~1+0=1	4x2=8
5x7=35~3+5=8	5x2=10~1+0=1
6x7=42~4+2=6	6x2=12~1+2=3
7x7=49~4+9=13~1+3=4	7x2=14~1+4=5
8x7=56~5+6=11~1+1=2	8x2=16~1+6=7
9x7=63~7+2=9	9x2=18~1+8=9

1x5=7	1x4=4
2x5=10~1+0=1	2x4=8
3x5=15~1+5=6	3x4=12~1+2=3
4x5=20~2+0=2	4x4=16~1+6=7
5x5=25~2+5=7	5x4=20~2+0=2
6x5=30~3+0=3	6x4=24~2+4=6
7x5=35~3+5=8	7x4=28~2+8=10~1+0=1
8x5=40~4+0=4	8x4=32~3+2=5
9x5=45~4+5=9	9x4=36~3+6=9

fig.11

This behavior is similar for the number images: 7 and 2, and 5 and 4, see (fig. 10 and 11).

Vortex Based Mathematics multiplication tables

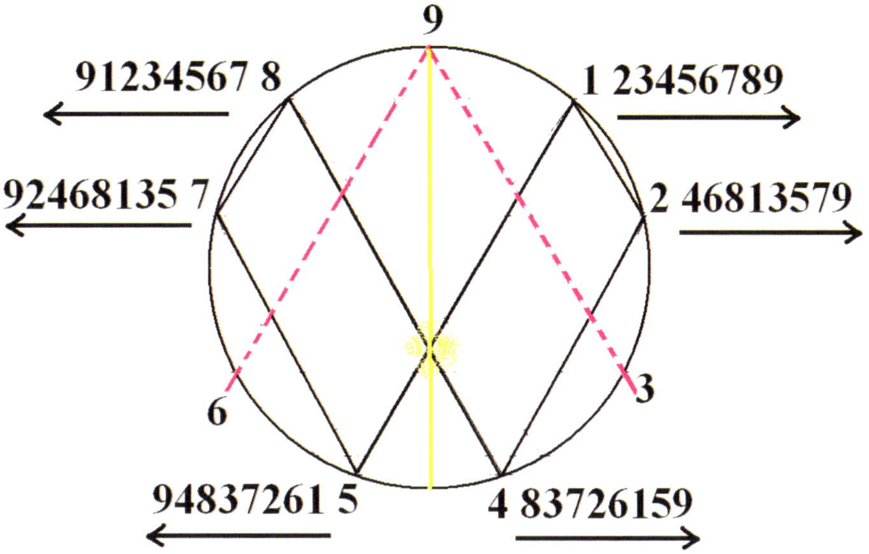

fig.12

1x8=8	1x1=1
2x8=7	2x1=2
3x8=6	3x1=3
4x8=5	4x1=4
5x8=4	5x1=5
6x8=3	6x1=6
7x8=2	7x1=7
8x8=1	8x1=8
9x9=9	9x1=9

1x7=7	1x2=2
2x7=5	2x2=4
3x7=3	3x2=6
4x7=1	4x2=8
5x7=8	5x2=1
6x7=6	6x2=3
7x7=4	7x2=5
8x7=2	8x2=7
9x7=9	9x2=9

1x5=5	1x4=4
2x5=1	2x4=8
3x5=6	3x4=3
4x5=2	4x4=7
5x5=7	5x4=2
6x5=3	6x4=6
7x5=8	7x4=1
8x5=4	8x4=5
9x5=9	9x4=9

1x9=9	1x9=9
2x9=18~1+8=9	2x9=9
3x9=27~2+7=9	3x9=9
4x9=36~3+6=9	4x9=9
5x9=45~4+5=9	5x9=9
6x9=54~5+4=9	6x9=9
7x9=63~6+3=9	7x9=9
8x9=72~7+2=9	8x9=9
9x9=81~8+1=9	9x9=9

fig.13

9

8 going down in increments of 1	**1 going up in increments of 1**
91234567 8	**1 23456789**
7 going down in increments of 2	**2 going up in increments of 2**
92468135 7	**2 46813579**
93693693 6	**3 69369369**
5 going down in increments of 4	**4 going up in increments of 4**
94837261 5	**4 83726159**

fig. 14 The number 1 is considered 10 in some cases.

In the (fig. 14) we can see that the number 1 increases in intervals of 1, and the number 8 is decreasing in intervals of 1. Thus, there is a polarity in the numbers that are images. When the number 1 is negative and increases in intervals of 1, the number 8 is positive decreasing in intervals of 1 or vice versa when the number 1 is positive and increases in intervals of 1, the number 8 is negative decreasing in intervals of 1. This happens with the image numbers 7 and 2 where the increments and decrements are at intervals of 2, and the image numbers 5 and 4 where the increments and decrements are at intervals of 4.

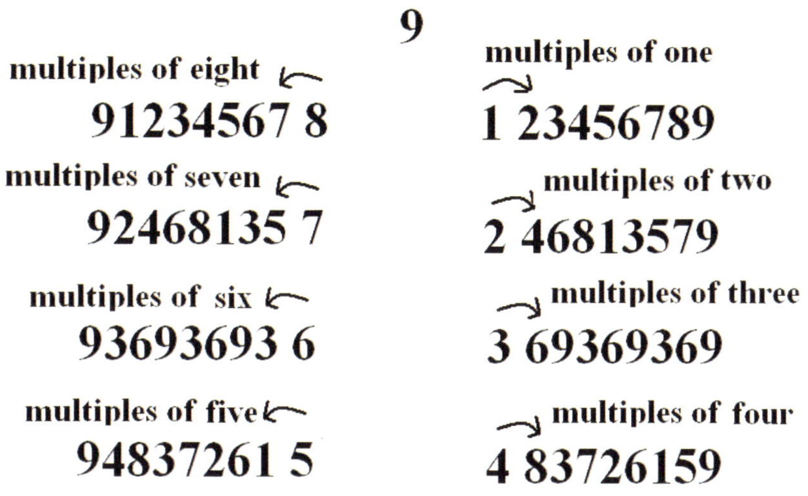

fig. 15 In the case of the numbers: 2,4,5 y 7 the number 1 is considered 10.

In the (fig. 15) the number 1 is increased in multiples of 1, and the number 8 increases in multiples of 8 and so happens with the numbers 2,3,4,5,6 and 7.

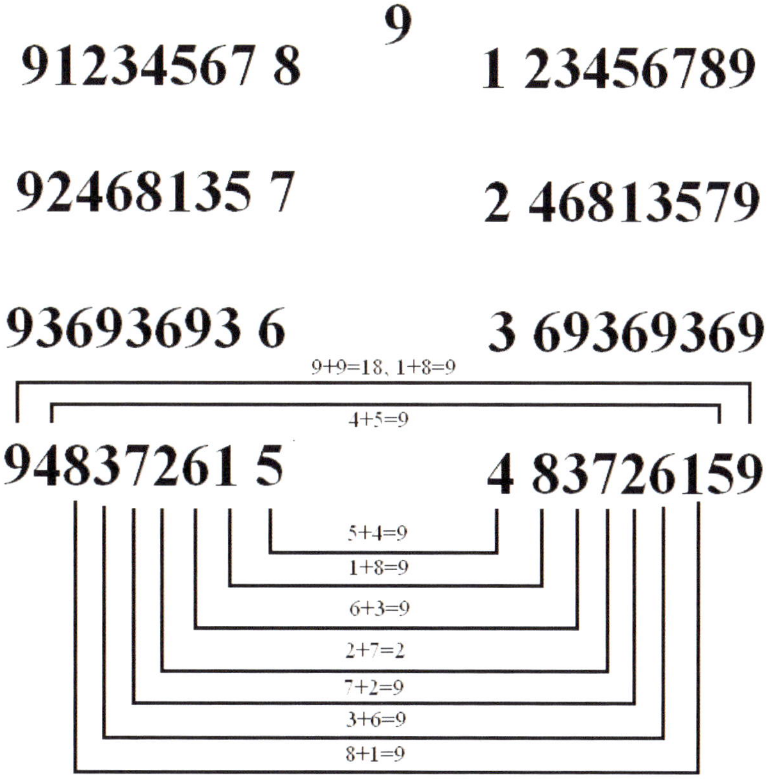

fig. 16

In (fig.16), the sum of each alignment of numbers image is equal to 9.

The multiplication function creates circuits that do not break, see (fig. 17 and 18).

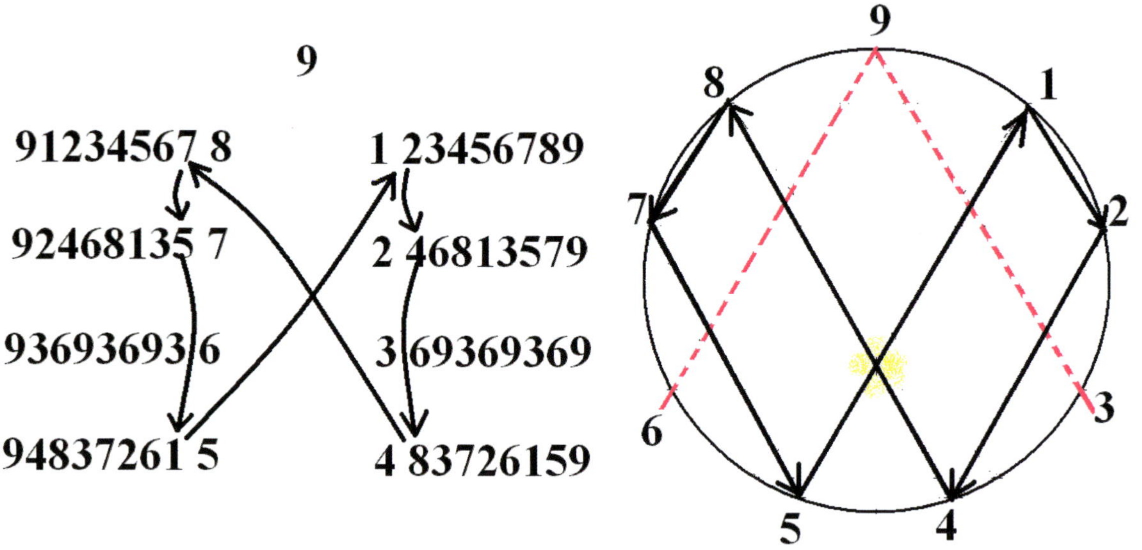

fig. 17

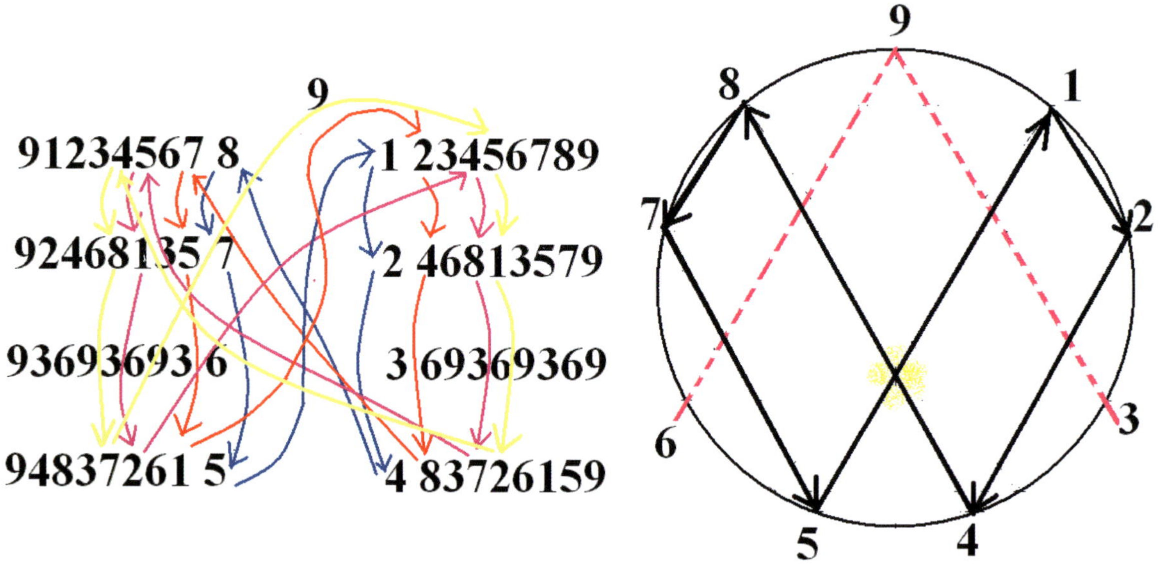

fig. 18

The numbers that exist in the world of creation are: 1,2,4,5,7 and 8, and they form the geometry of nature which is essentially hexagonal. The numbers 8 and 1, 7 and 2, 5 and 4 are the image numbers with positive or negative polarization.

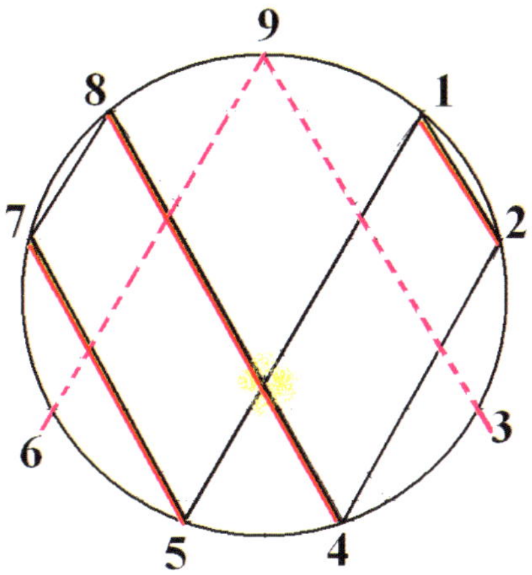

fig. 19

The number 9 splits up in numbers 3 and 6, see (fig. 19). We can observe that the alignments of the numbers 1 and 2, 8 and 4, 7 and 5 are parallel to the alignment of numbers 9 and 3, and the sum of these numbers are:
1+2=3
9+3=12 ~ 1+2=3
8+4=12 ~ 1+2=3
7+5=12 ~ 1+2=3

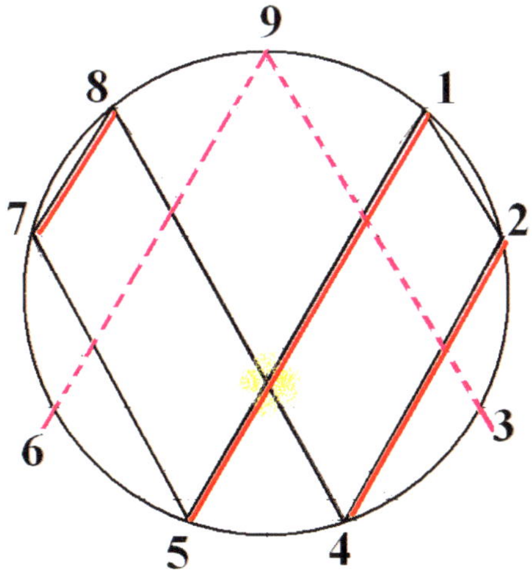

fig. 20

In (fig. 20) we can observe that alignments of the numbers 8 and 7, 1 and 5, 2 and 4 are parallel to the alignment of numbers 9 and 6, and the sum of these numbers is:
8+7=15 ~ 1+5=6
9+6=15 ~ 1+5=6
1+5=6
2+4=6

The doubling circuit follows a trajectory similar to the infinity symbol which oscillates infinitely from the number 3 to the number 6, and the number 6 to the number 3. The numbers 6 and 3 create the magnetic field in the doubling circuit. The doubling circuit is actually a vortex. The number 9 is the emanation linear* coming from the center of the vortex, see (fig. 21 and 22).

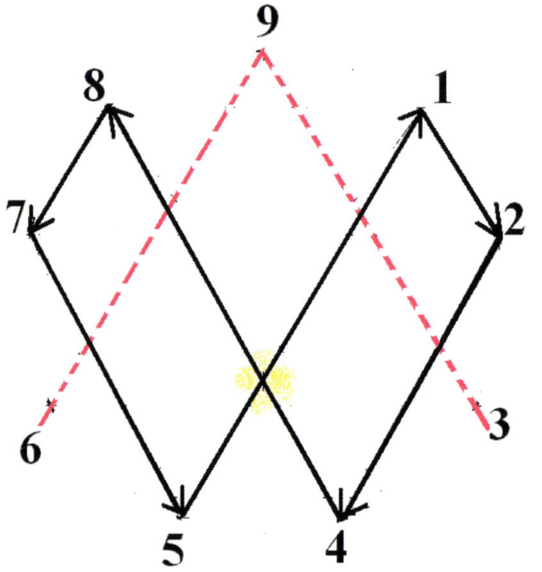

fig. 21

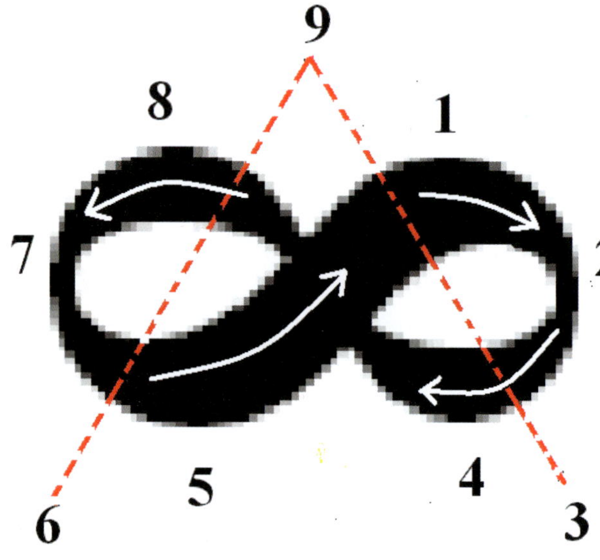

fig. 22

*Note: The concepts of Vortex Based Mathematics will be understood as the reader reads more about this subject, this book is presenting only the basic concepts.

Reciprocal numbers

Case 1. When the number 1 is the control, we multiply 1x1=1 and advance to the numbers 2 and 5. We can say that number 1 is reciprocal of itself.

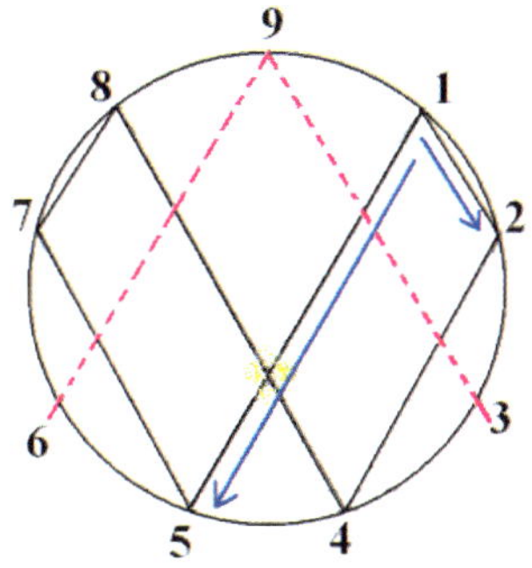

fig. 23

2x5=10~1+0=1

The numbers 2 and 5 are called reciprocals because the multiplication of both is equal to one, see (fig. 23). Now, we advance to the number 4 and 7.

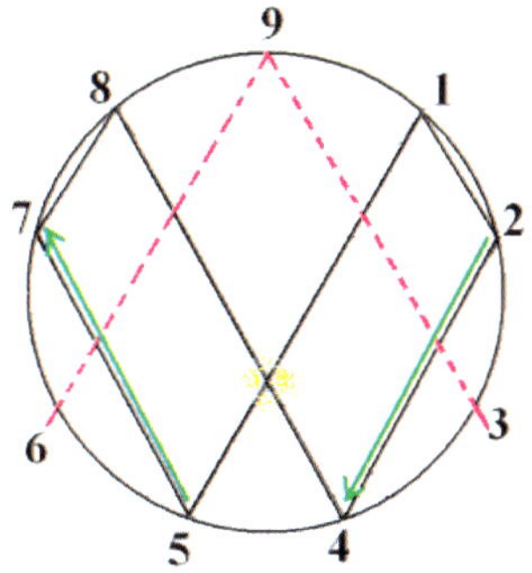

fig. 24

4x7=28~2+8=10~1+0=1

The numbers 4 and 7 are called reciprocals because the multiplication of both is equal to one, see (fig. 24). Now, we advance to the number 8 and 8.

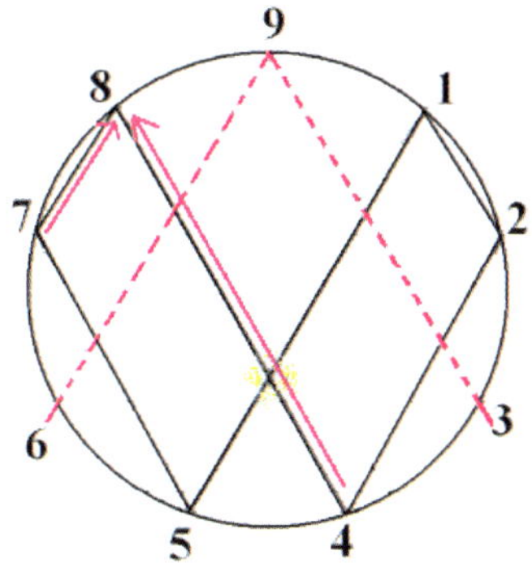

fig. 25

8x8=64~6+4=10~1+0=1

The numbers 8 and 8 are called reciprocals because the multiplication of both is equal to one, see (fig. 25). Number 8 is reciprocal of itself.

Case 2. When the number 2 is the control, we multiply 2x2=4 and advance to the numbers 1 and 4.

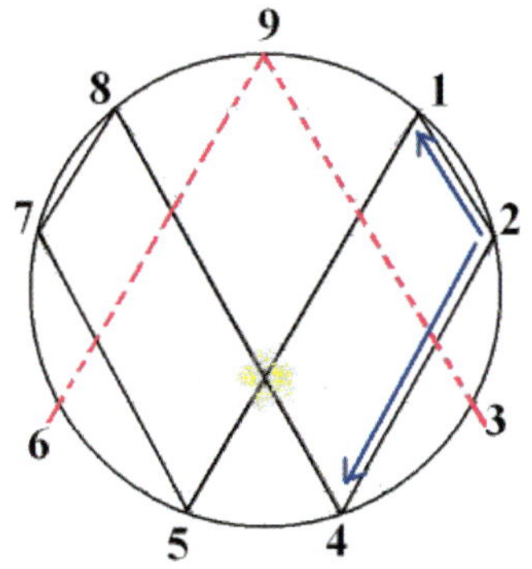

fig. 26

1x4=4

The numbers 1 and 4 are called reciprocals because the multiplication of both is equal to four, see (fig. 26). Now, we advance to the number 5 and 8.

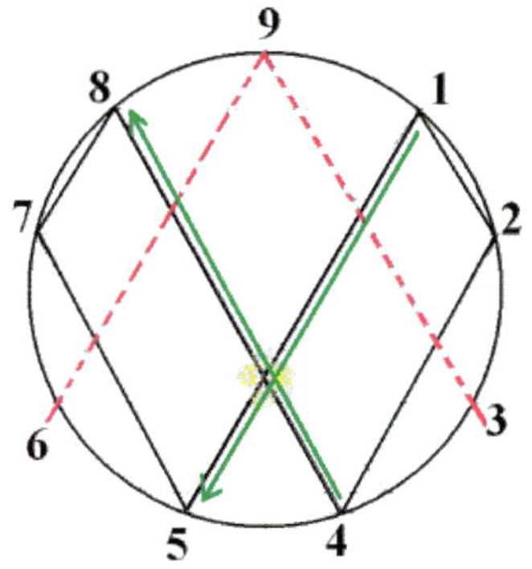

fig. 27

5x8=40~4+0=4

The numbers 5 and 8 are called reciprocals because the multiplication of both is equal to four, see (fig. 27). Now, we advance to the number 7 and 7.

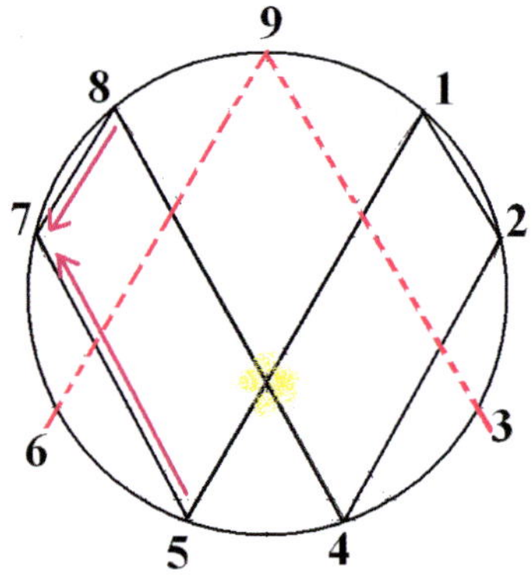

fig. 28

7x7=49~4+9=13~1+3=4

The numbers 7 and 7 are called reciprocals because the multiplication of both is equal to four, see (fig. 28). Number 7 is reciprocal of himself.

Case 3. When the number 4 is the control, we multiply 4x4=16 and 1 plus 6 is equal 7, see (fig. 28).

8x2=16~1+6=7

1x7=7

5x5=25~2+5=7

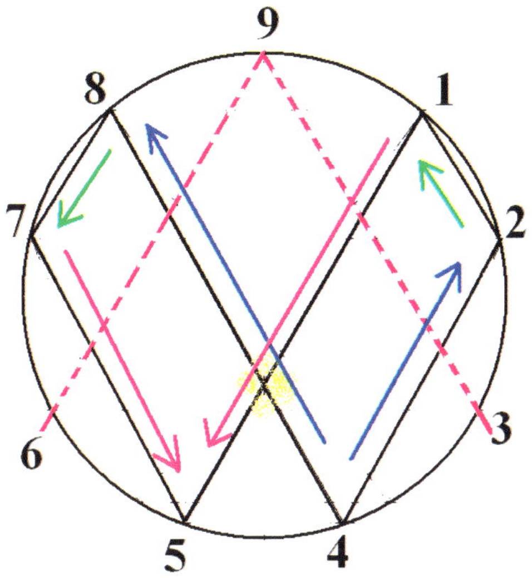

fig. 28

Case 4. When the number 5 is the control, we multiply 5x5=25 and 2 plus 5 is equal 7, see (fig. 29).

7x1=7

8x2=16~1+6=7

4x4=16~1+6=7

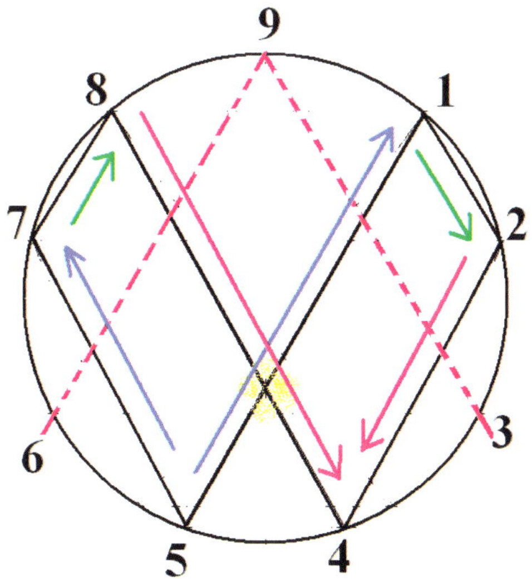

fig. 29

Case 5. When the number 7 is the control, we multiply 7x7=49 y 4+9=13 y 1+3=4, see (fig. 30).

8x5=40~4+0=4

4x1=4

2x2=4

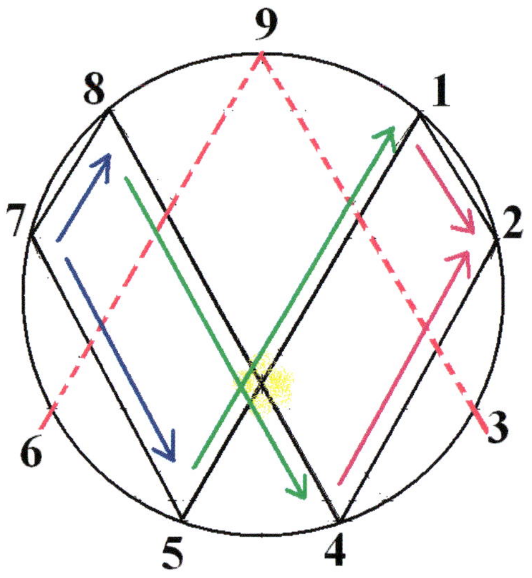

fig. 30

Caso 6. When the number 8 is the control, we multiply 8x8=64 and 6+4=10 and 1+0=1 see (fig. 31).

7x4=28~2+8=10~1+0=1

5x2=10~1+0=1

1x1=1

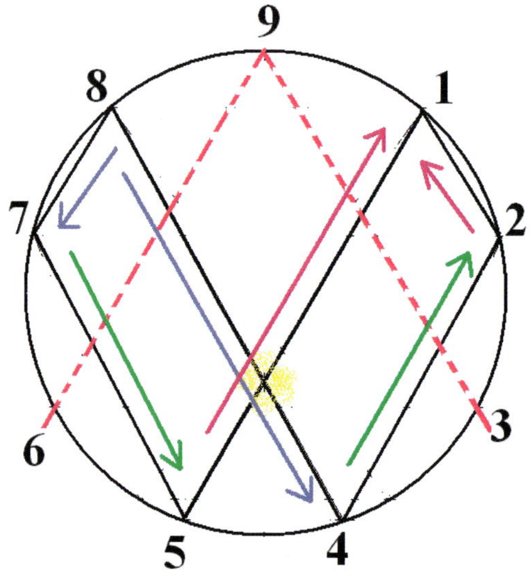

fig. 31

Reciprocal numbers are separated, but their multiplication always carries them to the number where they departed. As in the case 1, see (fig. 32).

1x1=1

2x5=10~1+0=1

4x7=28~2+8=10~1+0=1

8x8=64~6+4=10~1+0=1

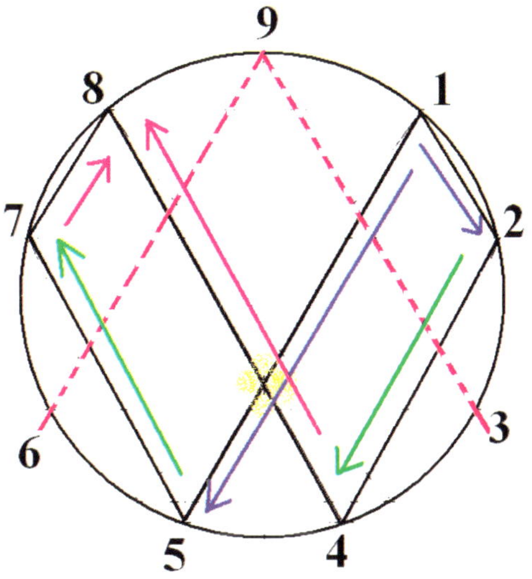

fig. 32

The reciprocal numbers in this case are: 1 and 1, 2 and 5, 4 and 7, 8 and 8. This principle is important in Vortex Based Mathematics, because it models the doubling circuits harmonic flow. In conventional mathematics, functions remain fixed and the numbers give them different values. In Vortex Based Mathematics occurs the opposite, the numbers remain static and the functions move, which is what most resembles reality. One can say that numbers are stationary vectors*. In Vortex Based Mathematics functions have the ability to model movement.

*Note: The concepts of Vortex Based Mathematics will be understood as the reader reads more about this subject, this book is presenting only the basic concepts.

Division

Dividing by any number is exactly the same as multiplying by its reciprocal.

In case 1, when the number 1 is the control, 2 and 5, 4 and 7 are reciprocals and numbers 1 and 8 are reciprocal of themselves.

Reciprocals 5 and 2:
Dividing by 5 is exactly the same as multiplying by 2, and vice versa dividing by 2 is exactly the same as multiplying by 5, see (fig. 33 and 34).

1x2=2	1/5=0.2=2
2x2=4	2/5=0.4=4
3x2=6	3/5=0.6=6
4x2=8	4/5=0.8=8
5x2=1	5/5=1
6x2=3	6/5=1.2=3
7x2=5	7/5=1.4=5
8x2=7	8/5=1.6=7
9x2=9	9/5=1.8=9

1/2=0.5=5	1x5=5
2/2=1	2x5=1
3/2=1.5=6	3x5=6
4/2=2	4x5=2
5/2=2.5=7	5x5=7
6/2=3	6x5=3
7/2=3.5=8	7x5=8
8/2=4	8x5=4
9/2=4.5=9	9x5=9

fig. 33

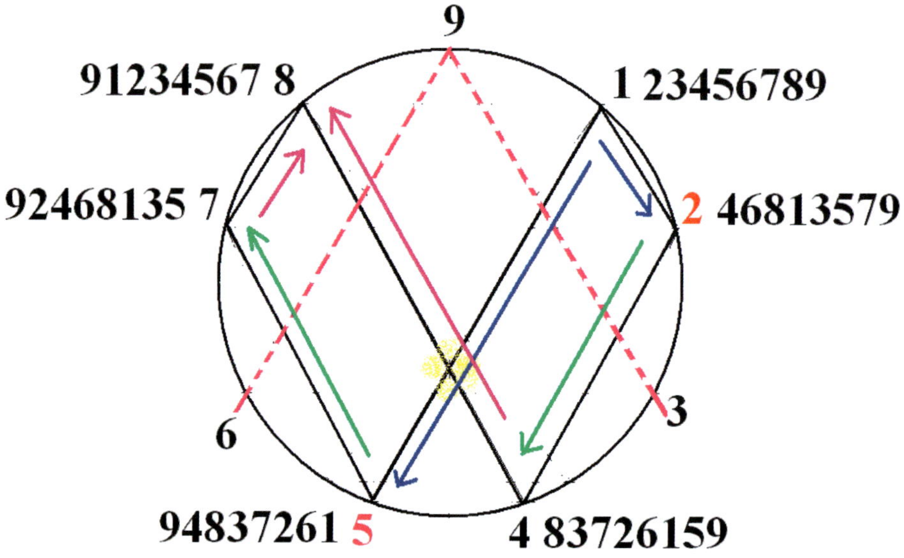

fig. 34

Reciprocals 7 and 4:
Dividing by 7 is exactly the same as multiplying by 4, and vice versa dividing by 4 is exactly the same as multiplying by 7. According to the principle of the reciprocal numbers, we can find out the actual values of: 1,2,3,4,5,6,7,8, and 9 divided by 7 even if this values appear to be a series of decimal numbers, see (fig. 35 and 36).

1x4=4	1/7=0.142857...=4
2x4=8	2/7=0.285714...=8
3x4=3	3/7=0.428571...=3
4x4=7	4/7=0.571428...=7
5x4=2	5/7=0.714285...=2
6x4=6	6/7=0.857142...=6
7x4=1	7/7=1
8x4=5	8/7=1.142857...=5
9x4=9	9/7=1.285714...=9

1/4=0.25=7	1x7=7
2/4=0.5=5	2x7=5
3/4=0.75=3	3x7=3
4/4=1	4x7=1
5/4=1.25=8	5x7=8
6/4=1.5=6	6x7=6
7/4=1.75=4	7x7=4
8/4=2	8x7=2
9/4=2.25=9	9x7=9

fig. 35

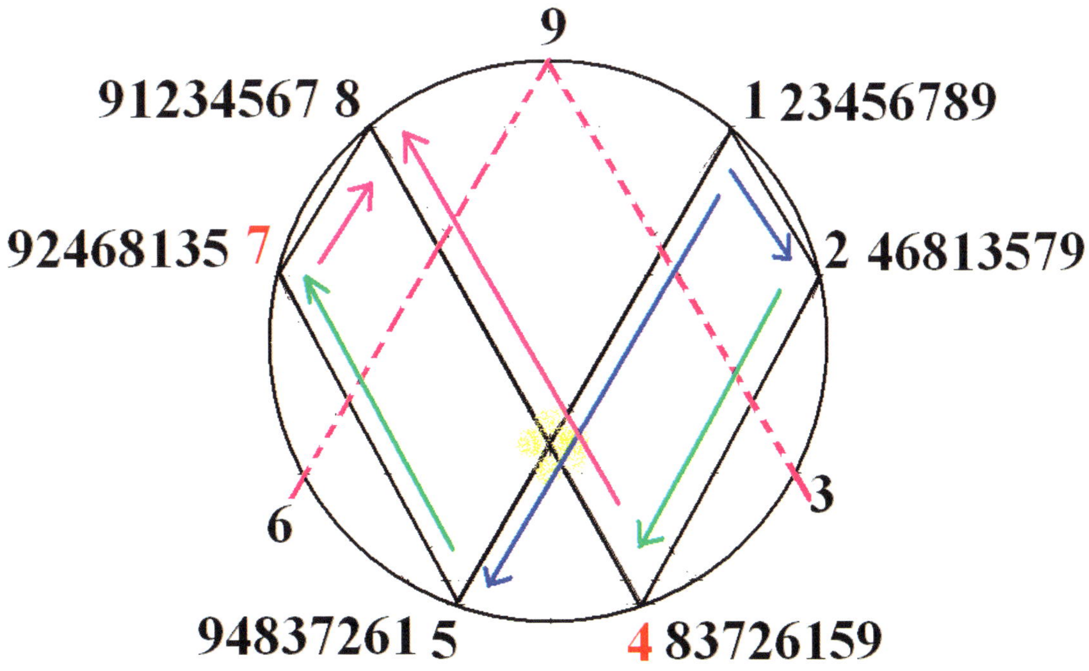

fig. 36

The number 1 is reciprocal of itself. Dividing the numbers: 1,2,3,4,5,6,7,8, and 9 by 1 is exactly the same as multiplying them by 1, see (fig. 37 and 38).

1x1=1	1/1=1
2x1=2	2/1=2
3x1=3	3/1=3
4x1=4	4/1=4
5x1=5	5/1=5
6x1=6	6/1=6
7x1=7	7/1=7
8x1=8	8/1=8
9x1=9	9/1=9

1x8=8	1/8=0.125=8
2x8=7	2/8=0.25=7
3x8=6	3/8=0.375=6
4x8=5	4/8=0.5=5
5x8=4	5/8=0.625=4
6x8=3	6/8=0.75=3
7x8=2	7/8=0.875=2
8x8=1	8/8=1
9x8=9	9/8=1.125=9

fig. 37

The number 8 is reciprocal of itself. Dividing the numbers: 1,2,3,4,5,6,7,8, and 9 by 8 is exactly the same as multiplying them by 8, see (fig. 37 and 38).

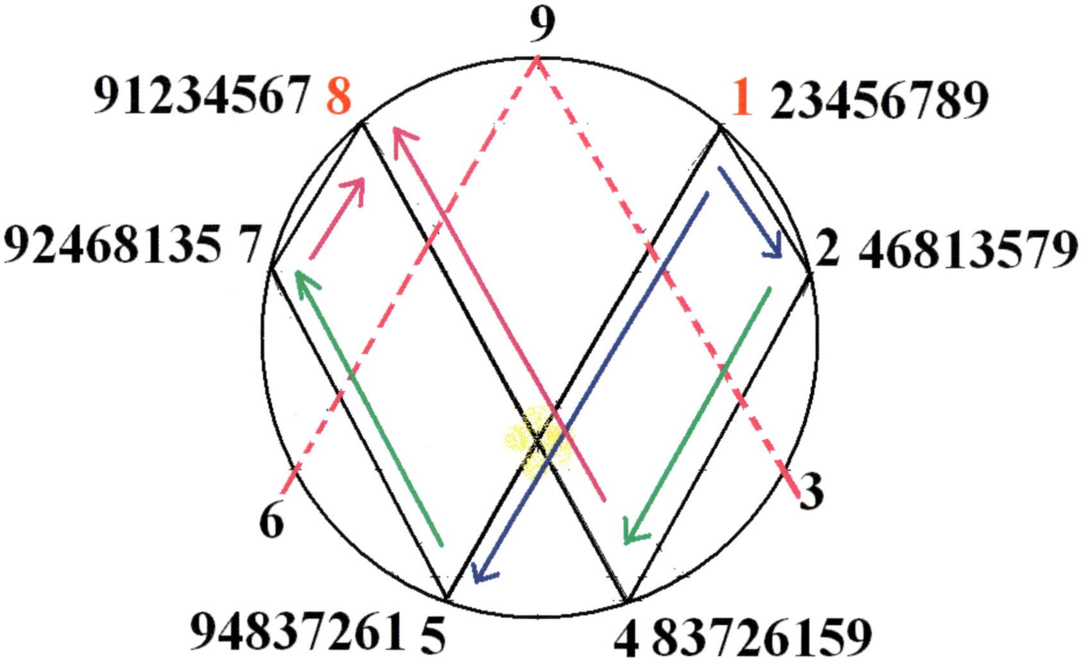

fig. 38

Dividing by 7

We know that 7 and 2 are image numbers, then dividing the numbers: 1,2,3,4,5,6,7,8, and 9 by 2 is the image of dividing 1,2,3,4,5,6,7,8, and 9 by 7, see (fig. 39 and 40).

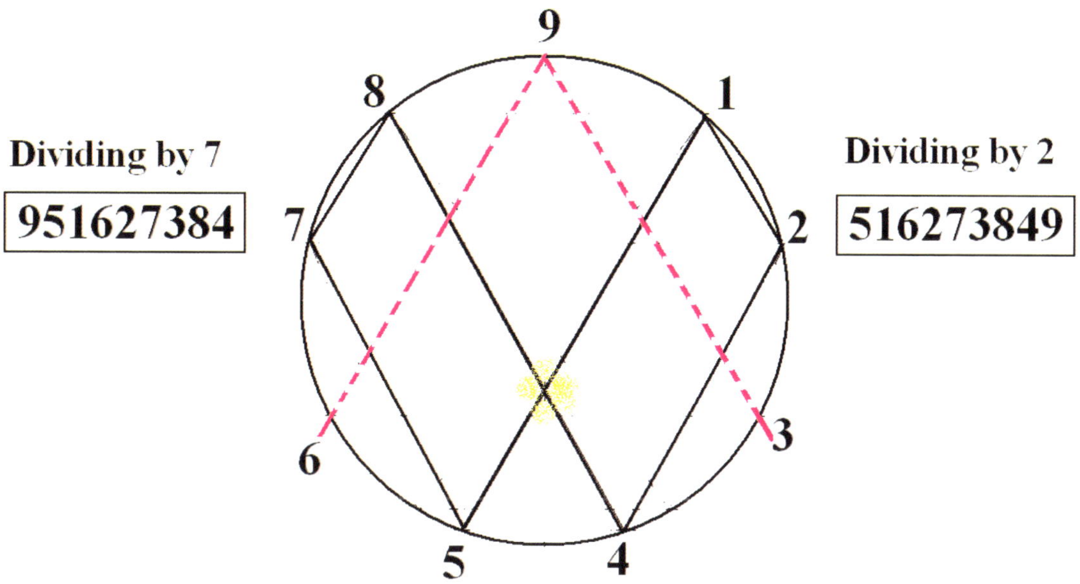

fig. 39

Dividing by 7	Dividing by 2
1/7=0.142857...=4	1/2=0.5=5
2/7=0.285714...=8	2/2=1
3/7=0.428571...=3	3/2=1.5=6
4/7=0.571428...=7	4/2=2
5/7=0.714285...=2	5/2=2.5=7
6/7=0.857142...=6	6/2=3
7/7=1	7/2=3.5=8
8/7=1.142857...=5	8/2=4
9/7=1.285714...=9	9/2=4.5=9

fig. 40

When we divide the numbers: 1,2,3,4,5,6,7,8, and 9 by 7, we get a series of repetitive decimal numbers corresponding to the doubling circuit numbers: 1,2,4,5,7, and 8, see (fig. 40).

1/7 = 0.142857142857….
2/7 = 0.28571422857142….
3/7 = 0.428571428571…..
4/7 = 0.5714228571….
5/7 = 0.714285714…
6/7 = 0.857142857…
7/7 = 1
9/7 = 1.285714286…
8/7 = 1.142857143……

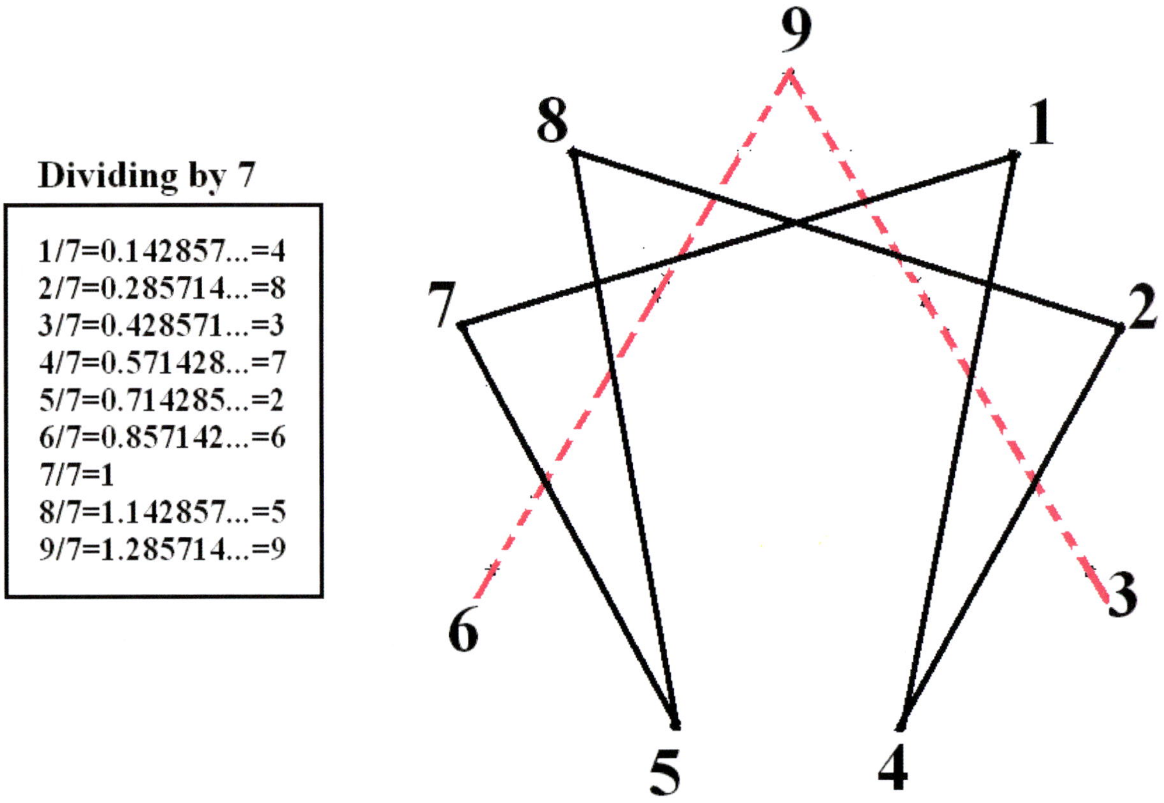

fig. 40 The enneagram dividing by 7

The division tables

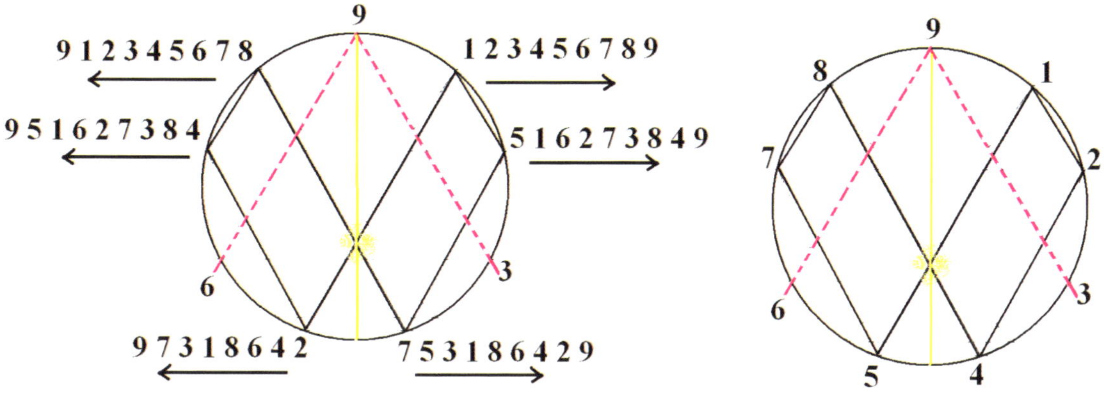

fig. 41

Dividing by 8
1/8=0.125=8
2/8=0.25=7
3/8=0.375=6
4/8=0.5=5
5/8=0.625=4
6/8=0.75=3
7/8=0.875=2
8/8=1
9/8=1.125=9

Dividing by 1
1/1=1
2/1=2
3/1=3
4/1=4
5/1=5
6/1=6
7/1=7
8/1=8
9/1=9

Dividing by 5
1/5=0.2=2
2/5=0.4=4
3/5=0.6=6
4/5=0.8=8
5/5=1
6/5=1.2=3
7/5=1.4=5
8/5=1.6=7
9/5=1.8=9

Dividing by 4
1/4=0.25=7
2/4=0.5=5
3/4=0.75=3
4/4=1
5/4=1.25=8
6/4=1.5=6
7/4=1.75=4
8/4=2
9/4=2.25=9

Dividing by 7
1/7=0.142857...=4
2/7=0.285714...=8
3/7=0.428571...=3
4/7=0.571428...=7
5/7=0.714285...=2
6/7=0.857142...=6
7/7=1
8/7=1.142857...=5
9/7=1.285714...=9

Dividing by 2
1/2=0.5=5
2/2=1
3/2=1.5=6
4/2=2
5/2=2.5=7
6/2=3
7/2=3.5=8
8/2=4
9/2=4.5=9

***Dividing by 3 and 6**

1/3=0.333333= 7	1/6=0.166666= 8
2/3=0.666666= 4	2/6=0.333333= 2
3/3=1	3/6=0.5= 5
4/3=1.333333= 8	4/6=0.666666= 4
5/3=1.666666= 5	5/6=0.833333= 7
6/3=2	6/6=1
7/3=2.333333= 9	7/6=1.166666= 9
8/3=2.666666= 6	8/6=1.333333= 3
9/3=3	9/6=1.5=6

fig. 41-a

The number 9

When the numbers: 1,2,3,4,5,6,7,8, and 9 are multiplied by 9 are equal to 9. When the number 9 is divided by the numbers: 12457, and 8 is equal to 9, but it is divided by 3, 6, and 9 is equal to 3, 6, and 1 respectively, see (fig. 41-b).

1x9=9	9/1=9
2x9=9	9/2=4.5=9
3x9=9	9/3=3
4x9=9	9/4=2.25=9
5x9=9	9/5=1.8=9
6x9=9	9/6=1.5=6
7x9=9	9/7=1.285714=9
8x9=9	9/8=1.125=9
9x9=9	9/9=1

fig. 41-b

*Note: The concepts of Vortex Based Mathematics will be understood as the reader reads more about this subject, this book is presenting only the basic concepts.

Addition and subtraction

Addition and subtraction functions in Vortex Based Mathematics are defined by the increase and decrease intervals in the number series, see (fig. 42).

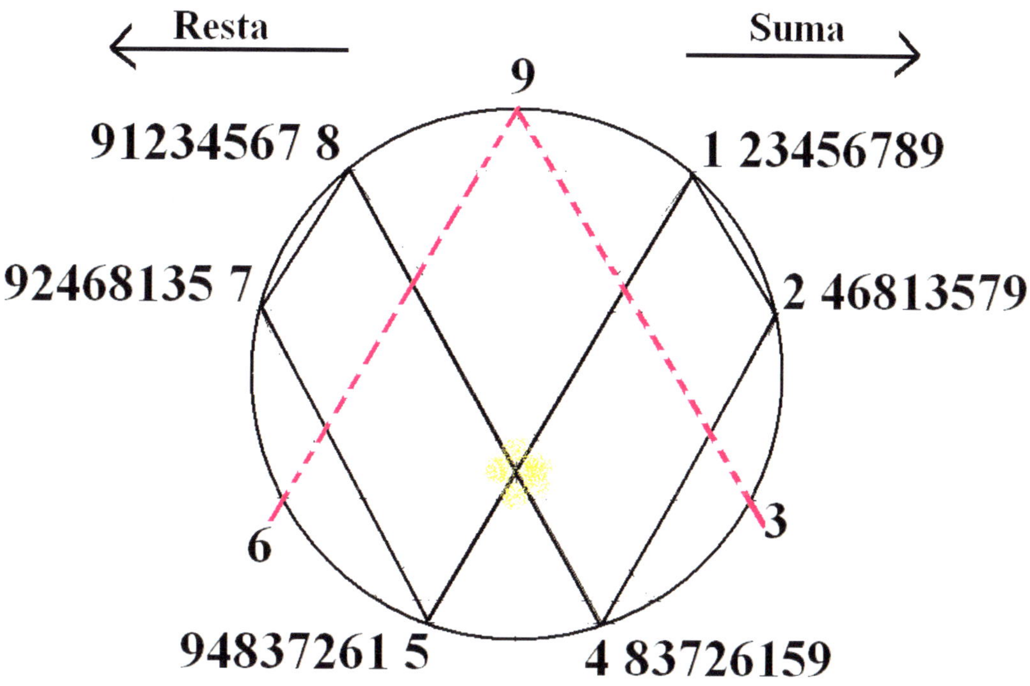

fig. 42

The family of numbers 1 4 7

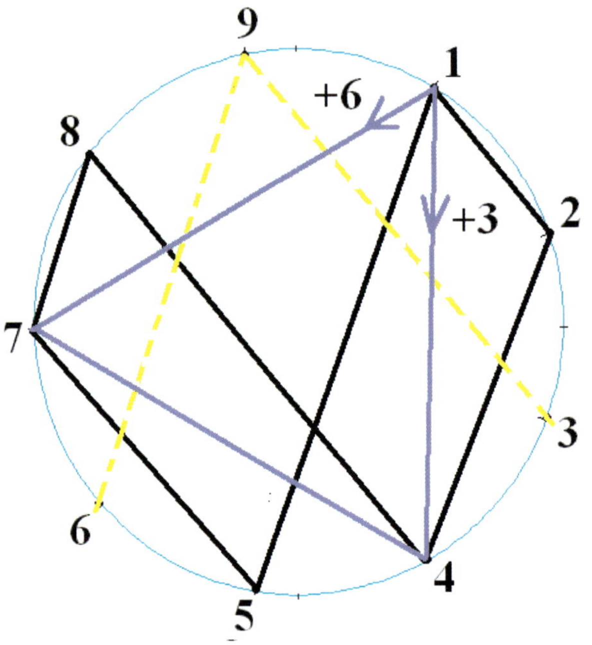

fig. 43

The number 3 and 6 are the doubling circuit's magnetic field. Starting on the number 1 and adding 3, 1+3=4, 4+3=7, 7+3=10~1+0=1, and we return to the number 1 which is the complete cycle to be repeated infinitely. In the same way, but in opposite direction we start on the number 1 and adding 6, 1+6=7, 7+6=13~1+3=4, 4+6=10~1+0=1, and we return to the number 1 which is the complete cycle to be repeated infinitely. This is the reason why the numbers 1 4 7 are members of the family, see (fig. 43 and 44).

1+3=4
4+3=7
7+3=10 ~ 1+0=1

10+3=13 ~ 1+3=4
13+3=16 ~ 1+6=7
16+3=19 ~ 1+9=1

....

1+6=7
7+6=13 ~ 1+3=4
13+6=19 ~ 1+9=10 ~ 1+0=1

19+6=25 ~ 2+5=7
25+6=31 ~ 3+1=4
31+6=37 ~ 3+7=10 ~ 1+0=1

....

fig. 44

The family of numbers 2 5 8

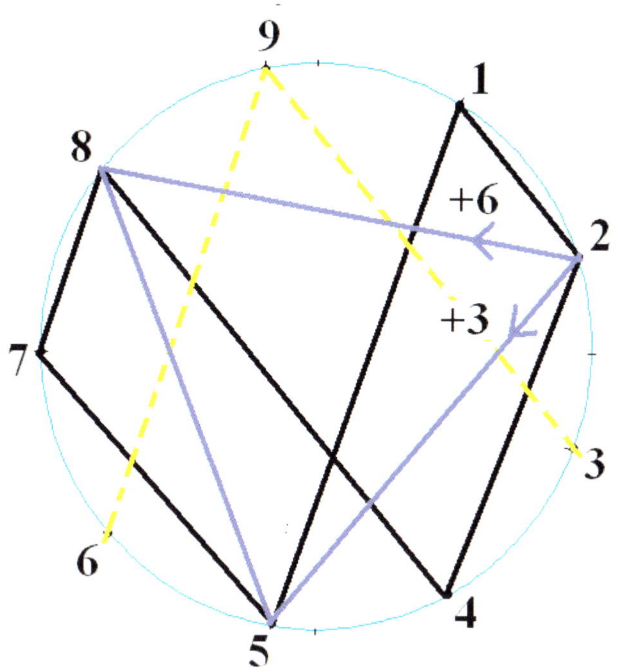

fig. 45

In the case of the family 2 5 8, Starting on the number 2 and adding 3, 2+3=5, 5+3=8, 8+3=11~1+1=2, and we return to the number 2 which is the complete cycle to be repeated infinitely. In the same way, but in opposite direction we start on the number 2 and adding 6, 2+6=8, 8+6=14~1+4=5, 5+6=11~1+1=2, and we return to the number 2 which is the complete cycle to be repeated infinitely. This is the reason why the numbers 2 5 8 are members of the family, see (fig. 45 and 46).

2+3=5
5+3=8
8+3=11 ~ 1+1=2

11+3=14 ~ 1+4=5
14+3=17 ~ 1+7=8
17+3=20 ~ 2+0=2
....

2+6=8
8+6=14 1+4=5
14+6=20 2+0=2

20+6=26 2+6=8
26+6=32 3+2=5
32+6=38 3+8=11 1+1=2
....

fig. 46

The family of numbers 9 3 6

This family includes the magnetic numbers 3 and 6, and the number 9 which is the linear emanation* coming from the center of the doubling circuit. Starting on the number 3 and adding 3, 3+3=6, 6+3=9, 9+3=12~1+2=3, and we return to the number 3 which is the complete cycle to be repeated infinitely, see (fig. 47).

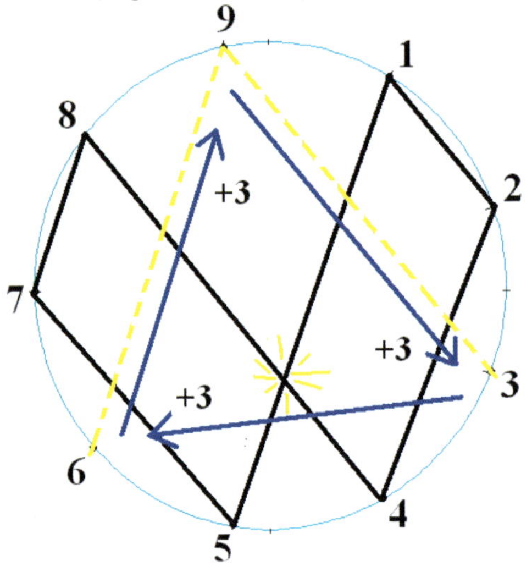

3+3=6
6+3=9
9+3=12   1+2=3

12+3=15   1+5=6
15+3=18   1+8=9
18+3=21   2+1=3
....

fig. 47

In the same way, but in opposite direction we start on the number 3 and adding 6, 3+6=9, 9+6=15~1+5=6, 6+6=12~1+2=3, and we return to the number 3 which is the complete cycle to be repeated infinitely, see (fig. 48).

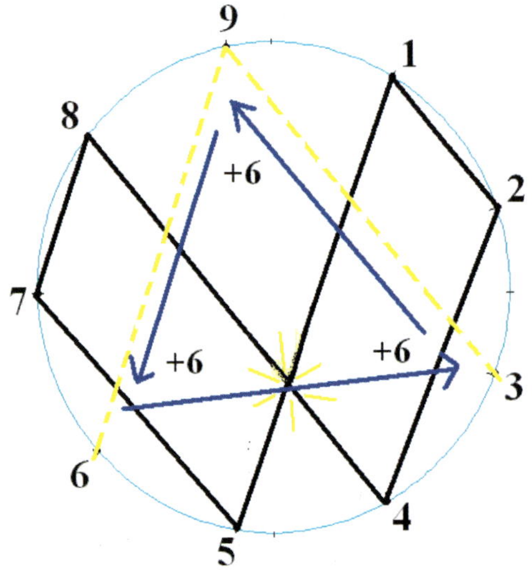

3+6=9
9+6=15   1+5=6
15+6=21   2+1=3

21+6=27   2+7=9
27+6=33   3+3=6
33+6=39   3+9=12
1+2=3
....

fig. 48

*Note: The concepts of Vortex Based Mathematics will be understood as the reader reads more about this subject, this book is presenting only the basic concepts.

In Vortex Based Mathematics, the numbers do not model geometry and arithmetic is not explaining the geometry. In this system, the numbers are the geometry. Thus, the number 3 is the vector moving forward in the form of a triangle, the number 6 is the vector moving in opposite direction giving the balance in an hexagonal shape*.

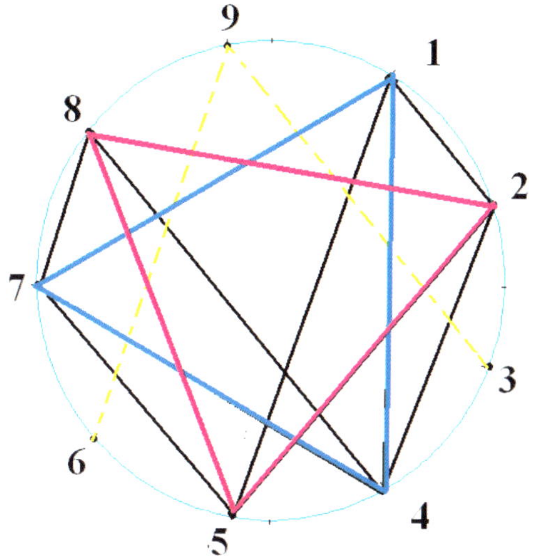

fig. 49 The groups of families: 1 4 7, 2 5 8, y 9 3 6

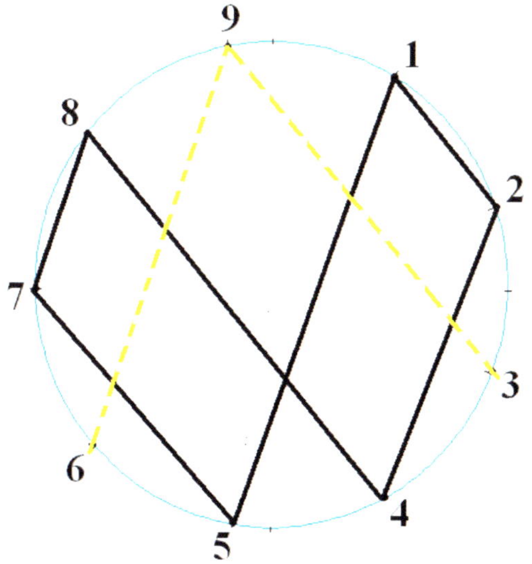

fig. 50 The doubling circuits and the numbers: 3, 6, and 9.

*Note: The concepts of Vortex Based Mathematics will be understood as the reader reads more about this subject, this book is presenting only the basic concepts.

The family of numbers 1 4 7 and 2 5 8

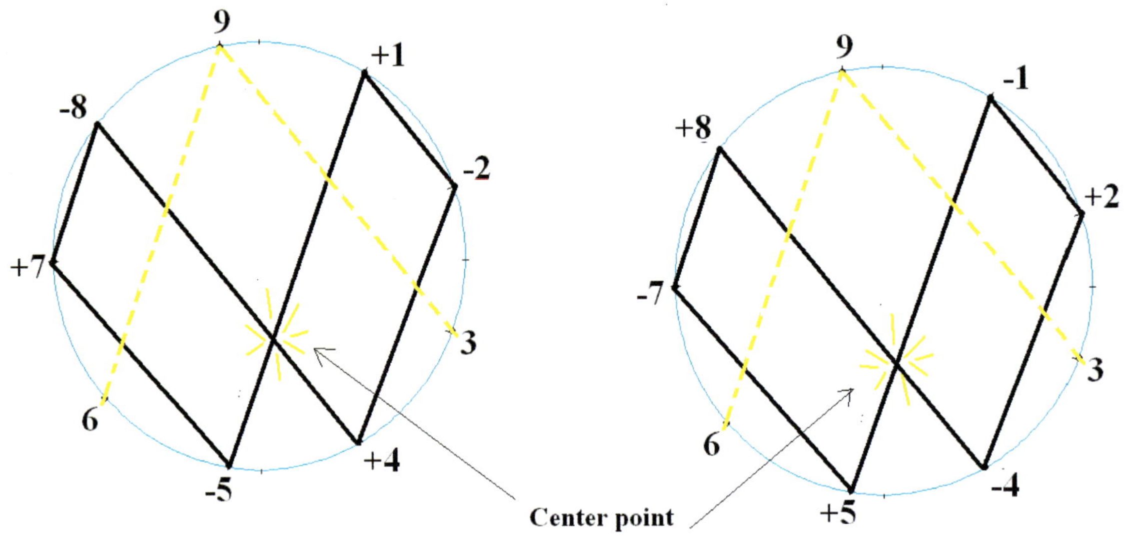

fig. 51

When the number 1 is positive, 8 is negative, 7 is positive, 5 is negative, 4 is positive, and 2 is negative or vice versa when number 1 is negative, 8 is positive, 7 is negative, 5 is positive, 4 is negative, and 2 is positive. The family numbers 1 4 7 and 2 5 8 have a positive or negative polarity, and they are determining the synchronization that exists in the doubling circuit. In Vortex Based Mathematics the doubling circuit represent the nature that is synchronized and controlled by the number 9 which is the linear emanation* coming from the center point, see (fig. 51).

*Note: The concepts of Vortex Based Mathematics will be understood as the reader reads more about this subject, this book is presenting only the basic concepts.

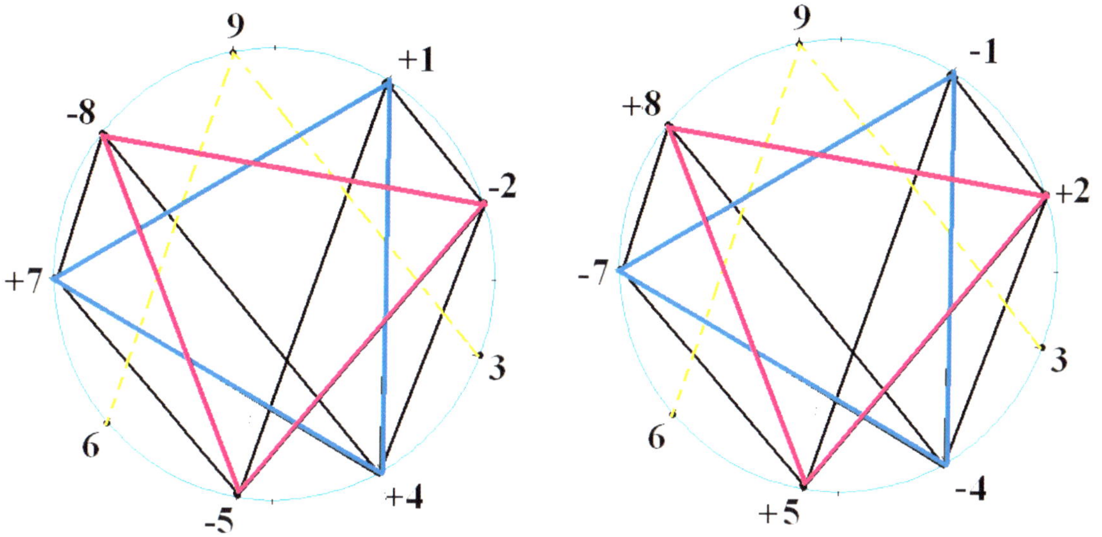

fig. 52

When the family numbers 1 7 4 is positive, the family numbers 2 8 5 is negative or vice versa when the family numbers 1 7 4 is negative, the family numbers 2 8 5 is positive, see (fig. 52). This means that there is a vector moving in one direction and another vector moving in the opposite direction. As we mentioned earlier, the number 3, 6, and 9 take the geometric shapes of a triangle, hexagon, and diamond* respectively, see (fig. 53).

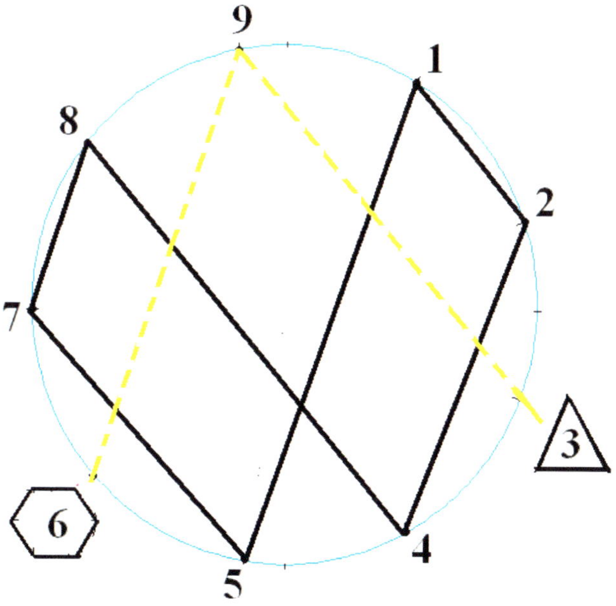

fig. 53

The family of numbers 1 4 7, 2 5 8, and 3 9 6 are separated by three and six. These numbers are activated at the same time. This activation or pulse produces a “flip-flop” or change of polarity in the doubling circuit. Thus, we can say that the doubling circuit is being activated in a binary way and in triplets. The doubling circuit has these triplets in the family numbers, but also has the binary property in the numbers of the doubling circuit: 1,2,4,8,7, and 5. If the family numbers 1 4 5 is positive, and the family numbers 2 5 8 is negative, the polarity of the numbers in the doubling circuit are: +1 -2 +4 -8 +7 -5. If the family numbers 1 4 5 is negative, and the family numbers 2 5 8 is positive, the polarity of the numbers in the doubling circuit are: -1 +2 -4 +8 -7 +5, see (fig. 54). This means that this system does not has only 9 numbers, but 18 numbers. In this system, each number has the property to be positively or negatively polarized at some point.

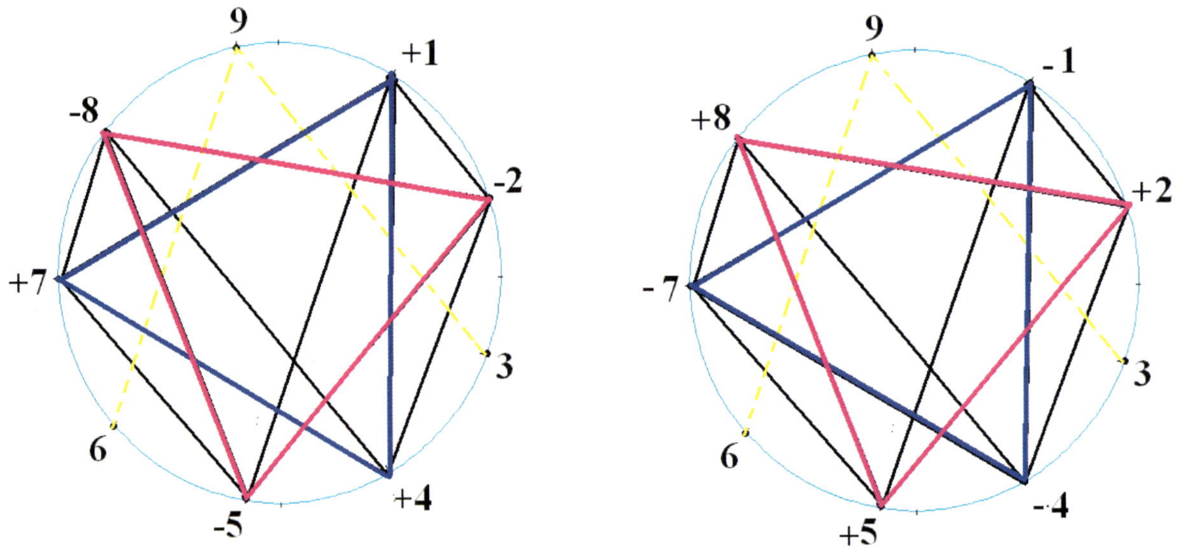

fig. 54

The numbers 3 6 9

In this system the number 9 is the initial emission or energy boost. When the number 9 is positive, the numbers 6 and 3 are negative. In other words, when the energy boost is emanated from the center of the doubling circuit, the magnetism is aimed at the center. As well as, when the impulse energy is absent, i.e. when the number 9 is negative, the numbers 3 and 6 are positive, see (fig. 55).

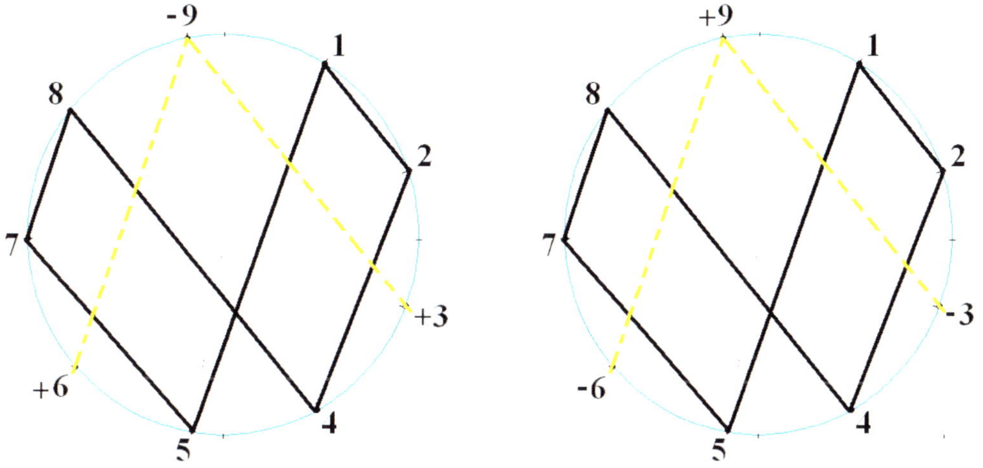

fig. 55

When the number 3 is positive, the number 9 is negative and the number 6 is positive. At this point it is when the flip-flop or change of polarity occurs, the number 6 positive changes to 6 negative and the cycle continues in the opposite direction, now the number 9 is positive and the number 3 is negative. The flip-flop occurs again, the number 3 negative changes to 3 positive, the number 9 is negative and the number 6 is positive, and this cycle is continuing infinitely, i.e. +3 -9 +6, -6 +9 -3, +3 -9 +6. What we are showing here is a pulse, the number 9 is a pulse o emanation*. In Vortex Based Mathematics this emanation is also called inertia-ether*. The number 9 is the pulse or emanation of energy which is causing the synchronized motion in the doubling circuit. The family numbers: 1 4 7, 2 5 8 y 3 9 6 are activated positively or negatively, this is the reason why there are 18 numbers in this system. If only three numbers are activated positively, in a certain period of time, means that 3/18 =1/6 of the numbers will be positive in certain period of time, see (fig.56).

*Note: The concepts of Vortex Based Mathematics will be understood as the reader reads more about this subject, this book is presenting only the basic concepts

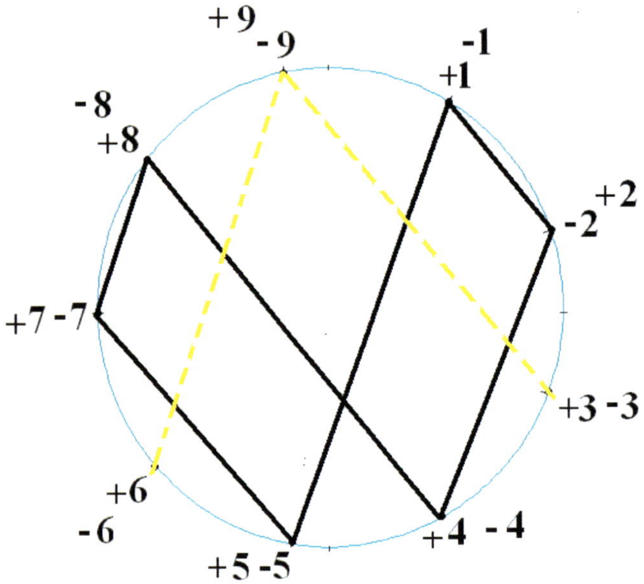

fig. 56 This is the representation of the doubling circuit in the two dimension model, but it shows what happens in the three dimension model.

The doubling circuit is a vortex which has an infinite path. In the center of this vortex is located the zero point. The function of the zero point is to compress by the upper part of the vortex and decompress at its bottom. There is an implosion in the upper part of the vortex and an explosion in its bottom, see (fig. 57).

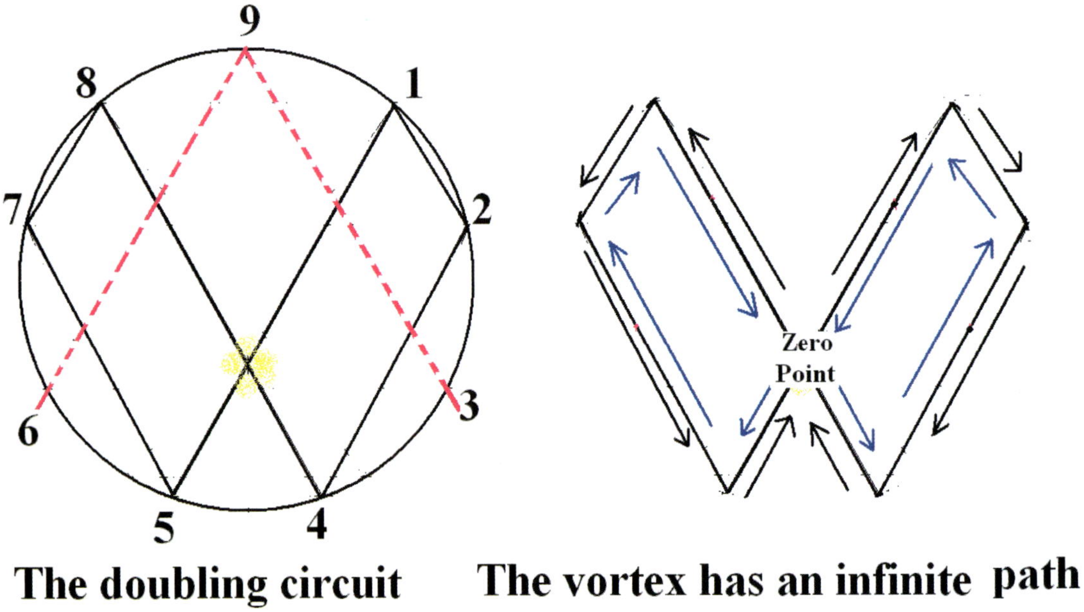

fig. 57

The doubling circuit never breaks

The multiplication functions are represented in horizontal and circular paths, these two conditions must be fulfilled to have a real system. This property creates the binary condition in the doubling circuit that never breaks, see (fig. 58). The multiplication series 1 and 2 are shown only to display this property.

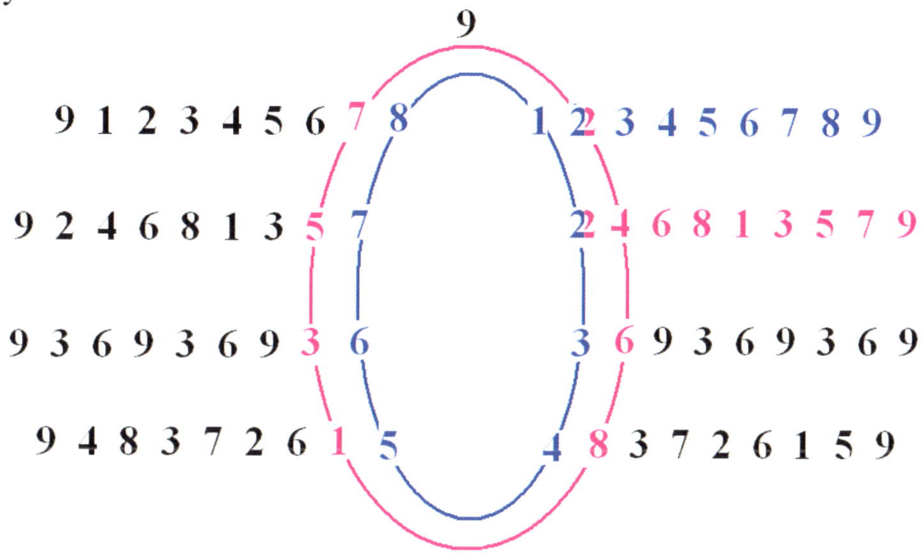

1x1=1	1x2=2	1x3=3	1x4=4
2x1=2	2x2=4	2x3=6	2x4=8
3x1=3	3x2=6	3x3=9	3x4=3
4x1=4	4x2=8	4x3=3	4x4=7
5x1=5	5x2=1	5x3=6	5x4=2
6x1=6	6x2=3	6x3=9	6x4=6
7x1=7	7x2=5	7x3=3	7x4=1
8x1=8	8x2=7	8x3=6	8x4=5
9x1=9	9x2=9	9x3=9	9x4=9

1x5=5	1x6=6	1x7=7	1x8=8
2x5=1	2x6=3	2x7=5	2x8=7
3x5=6	3x6=9	3x7=3	3x8=6
4x5=2	4x6=6	4x7=1	4x8=5
5x5=7	5x6=3	5x7=8	5x8=4
6x5=3	6x6=9	6x7=6	6x8=3
7x5=8	7x6=6	7x7=4	7x8=2
8x5=4	8x6=3	8x7=2	8x8=1
9x5=9	9x6=9	9x7=9	9x9=9

fig. 58

In the following cases the doubling circuit does not break, see (fig. 59).

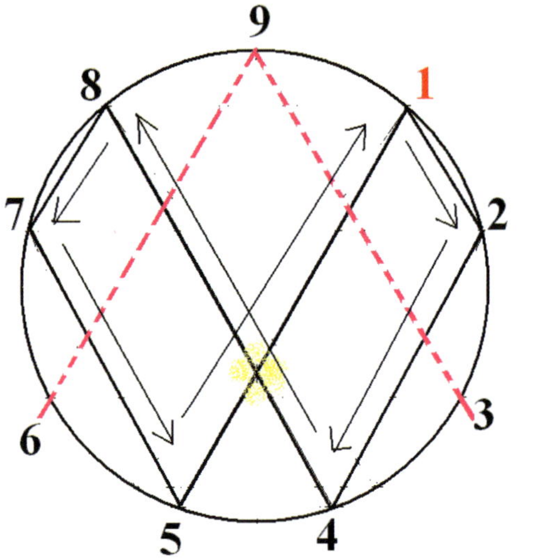

Control 1. Multiples of 1: 123456789

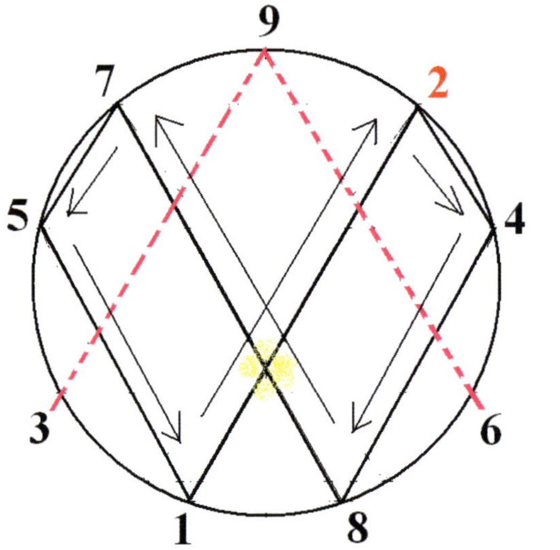

Control 2. Multiples of 2: 246813579

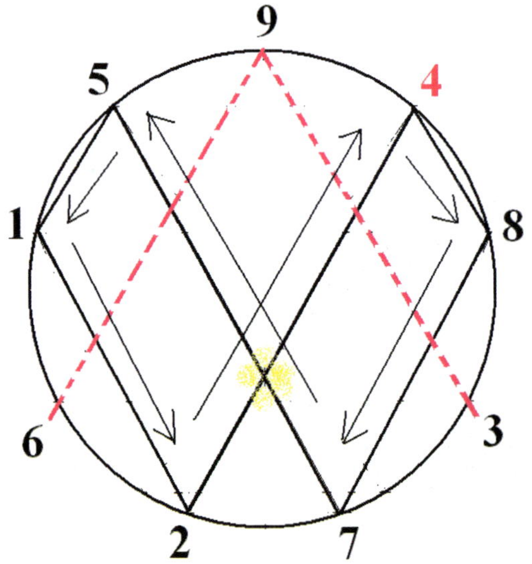

Control 4. Multiples of 4: 483726159

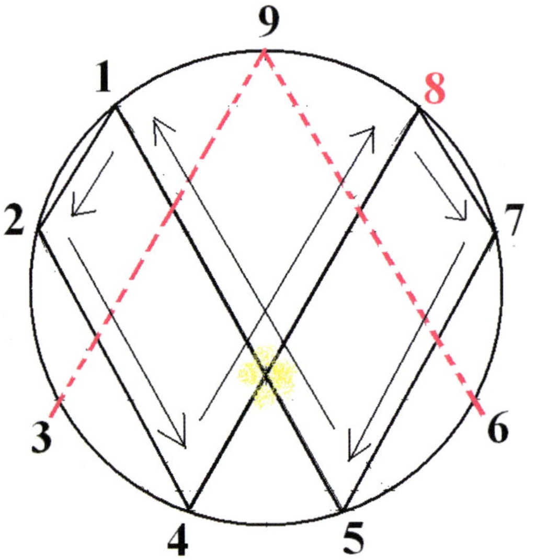

Control 8. Multiples of 8: 876543219

fig. 59

In the following examples, the doubling circuit breaks. We have the doubling circuit's numbers: 1,2,4,8,7, and 5, the numbers: 6 9 3, the polar images: 5 4, 1 8 and 2 7, and we can model the family numbers: : 147, 258 and 369, but the multiplication series are not valid as shown in (fig. 58).

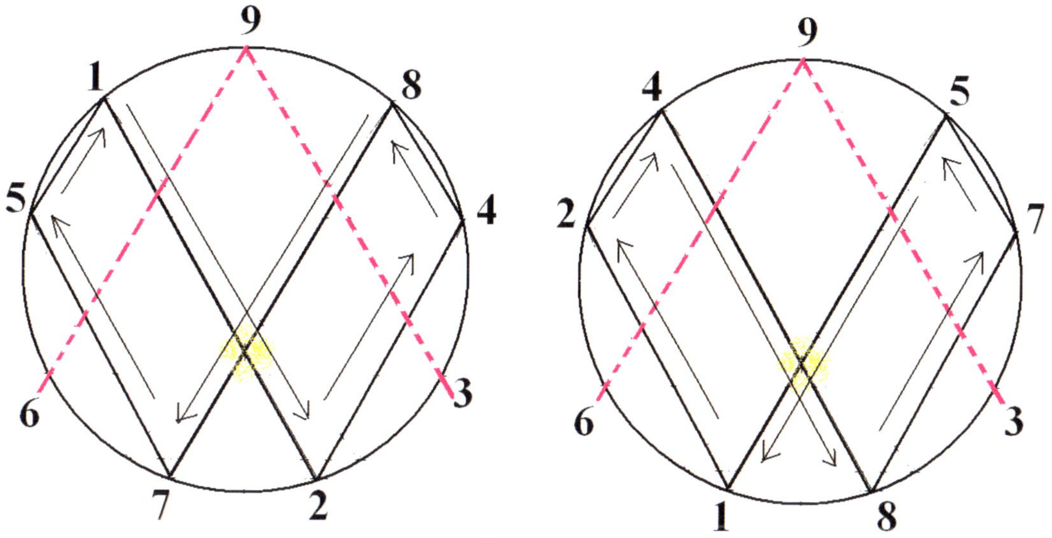

fig. 60

The (fig. 61, 62 and 63) shows how the doubling circuit is modeling motion. This motion is directed through the top of the vortex as an implosion and inverted in the center of the vortex like an explosion.

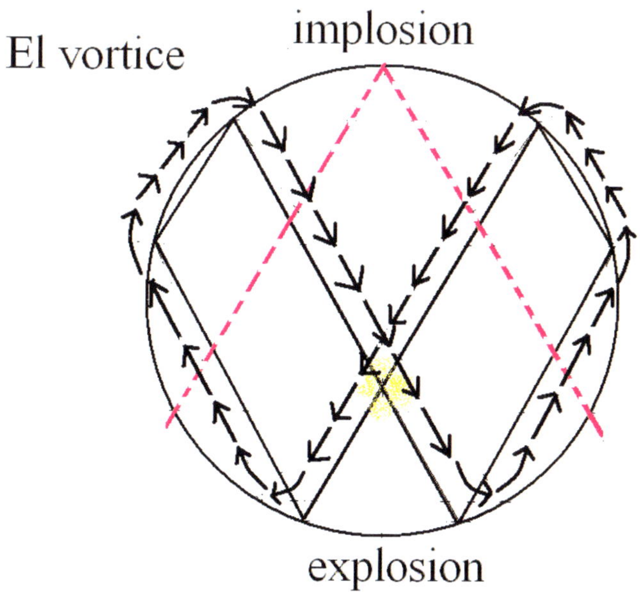

fig. 61

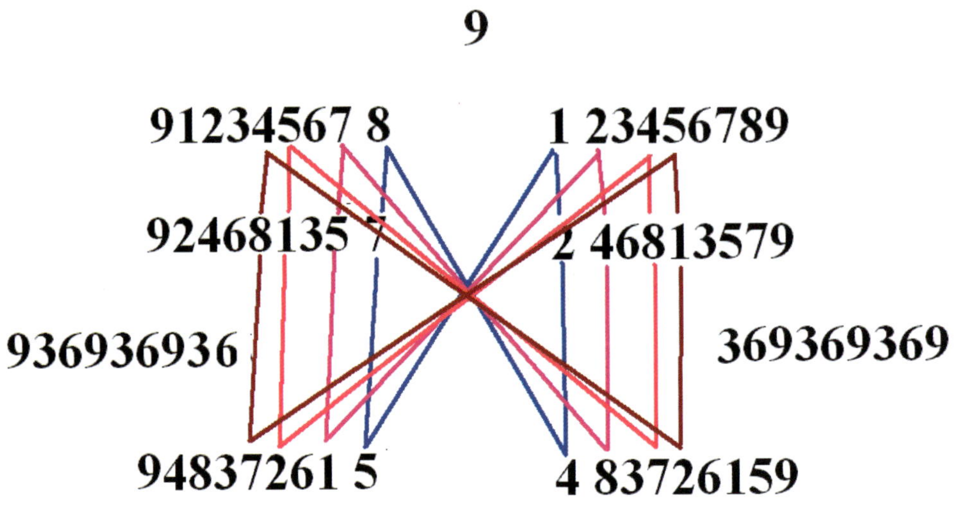

fig. 62

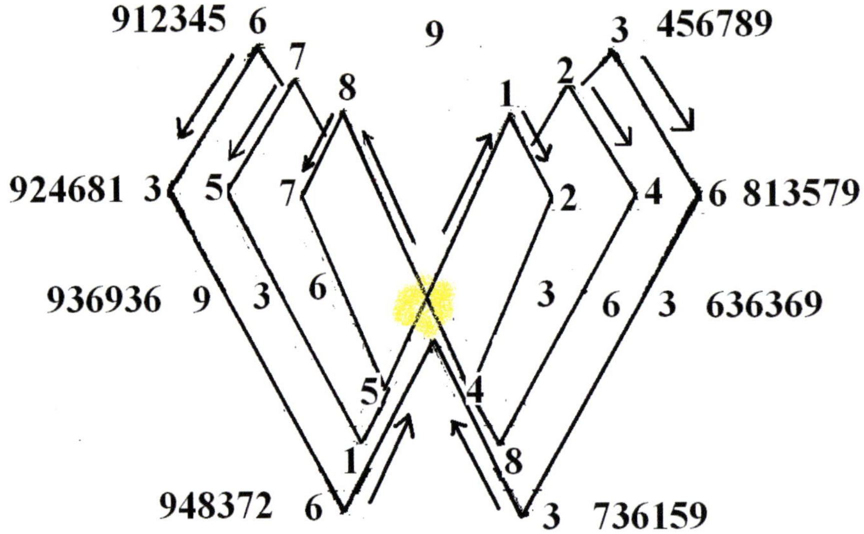

fig. 63

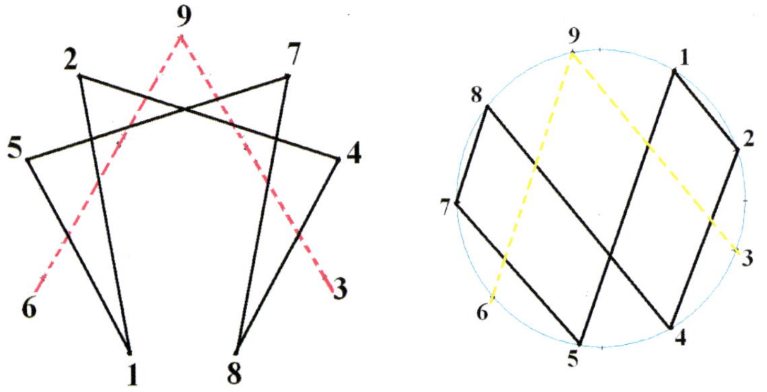

fig. 64 The doubling circuit

The power of the number 10

The exponents of the number 10 are functions of the division of the numbers: 1,2,4,5,7, and 8 by 2. Also, these functions are related to the multiplication of these numbers by 2, the reciprocal numbers, and the family numbers. Remember when we divided the numbers: 1,2,4,5,7, and 8 by 2, see (fig. 65).

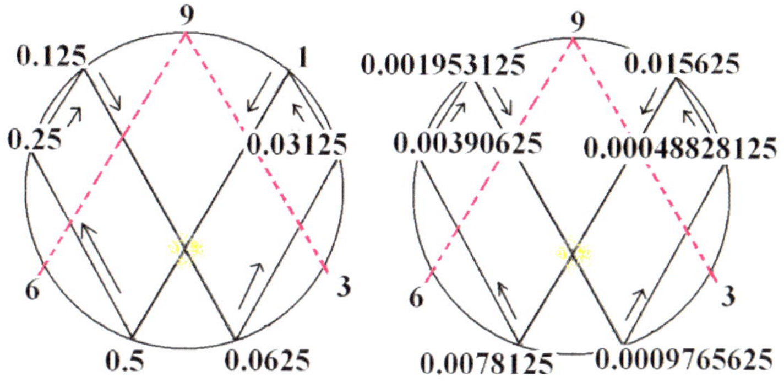

fig. 65

And its representation in fraction is, see (fig. 66).

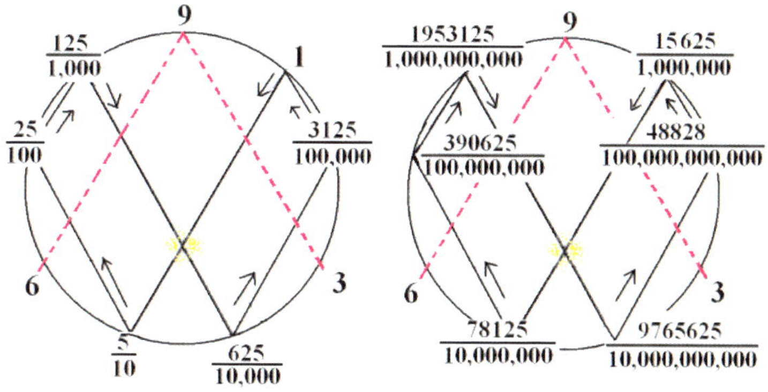

fig. 66

Thus the potential of the number 10 can be deduced, see (fig. 67).

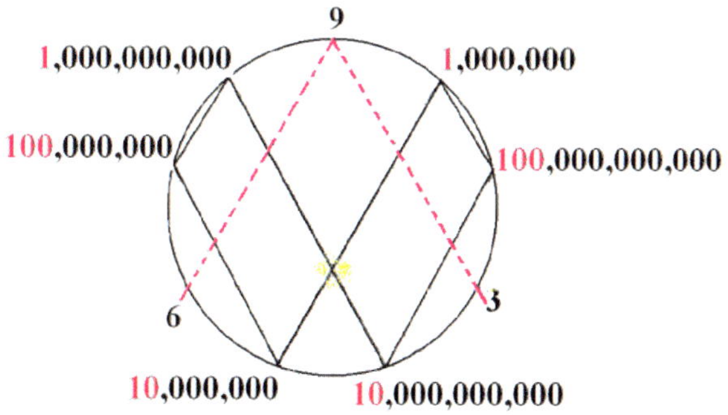

fig. 67

Quantum Numerology

We have presented the principles of the doubling circuit in detail. These principles will model the vortex* in the third dimension. This means that the numbers: 1,2,3,4,5,6,7,8, and 9 are going to have a space value. These numbers can be quantized in different geometric shapes as triangle, hexagon, and diamond. This book will show how quantizes them in a diamond shape see (fig. 68 and 69).

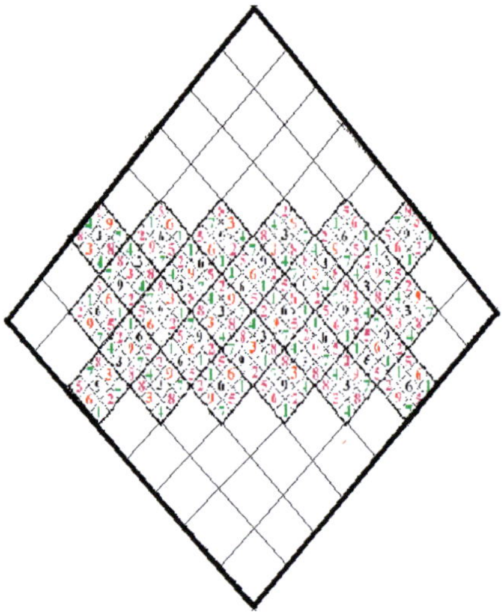

fig. 68

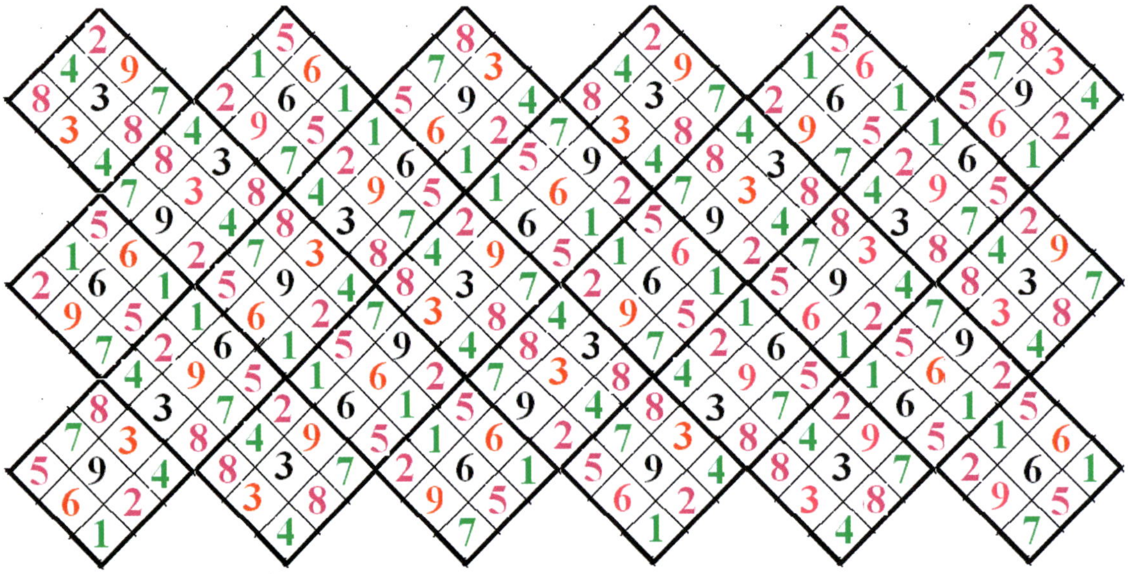

fig. 69

The quantum numerology map

This map quantizes or give a space value to the numbers: 1,2,3,4,5,6,7,8, and 9. Each tile or diamond on this map has a number, a function, a quality, and quantity, see (fig. 70).

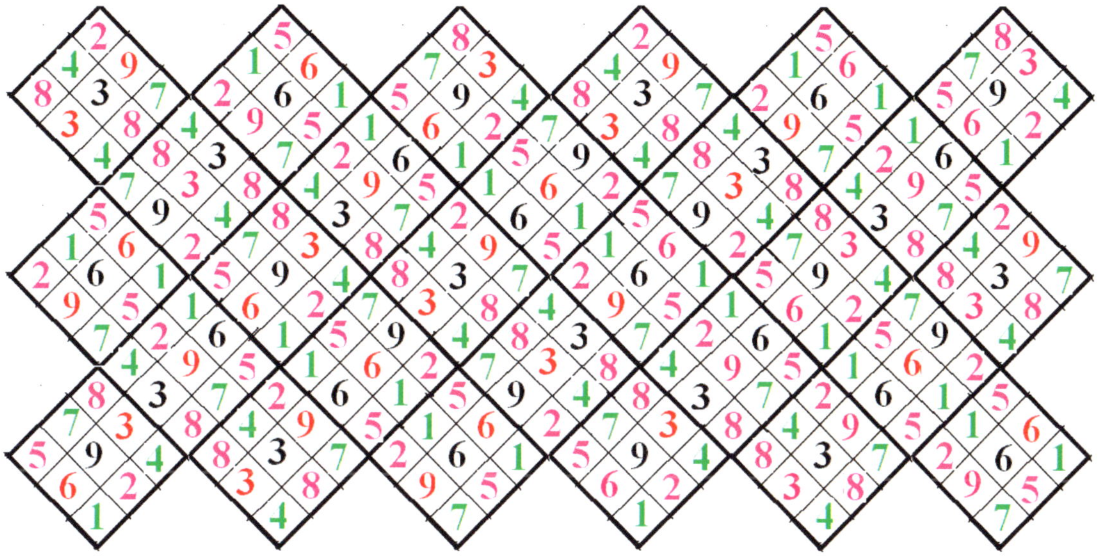

Family numbers: 1 4 7, 2 5 8,+3+9+6, -3 -9 -6

fig. 70

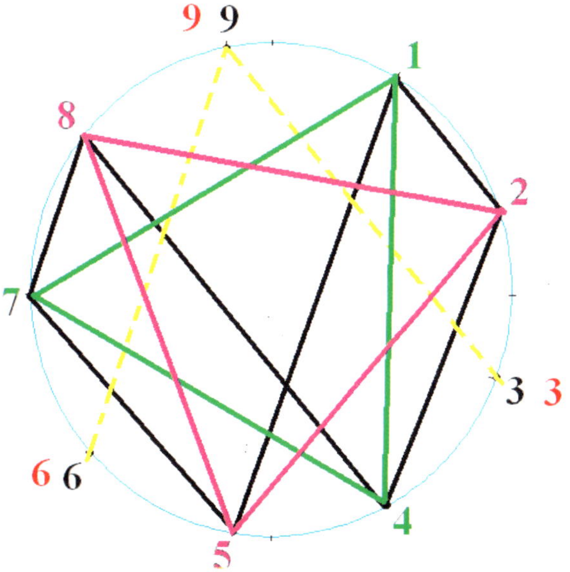

The family of numbers: 1 4 7, 2 5 8, 3 9 6, 3 9 6

fig. 71

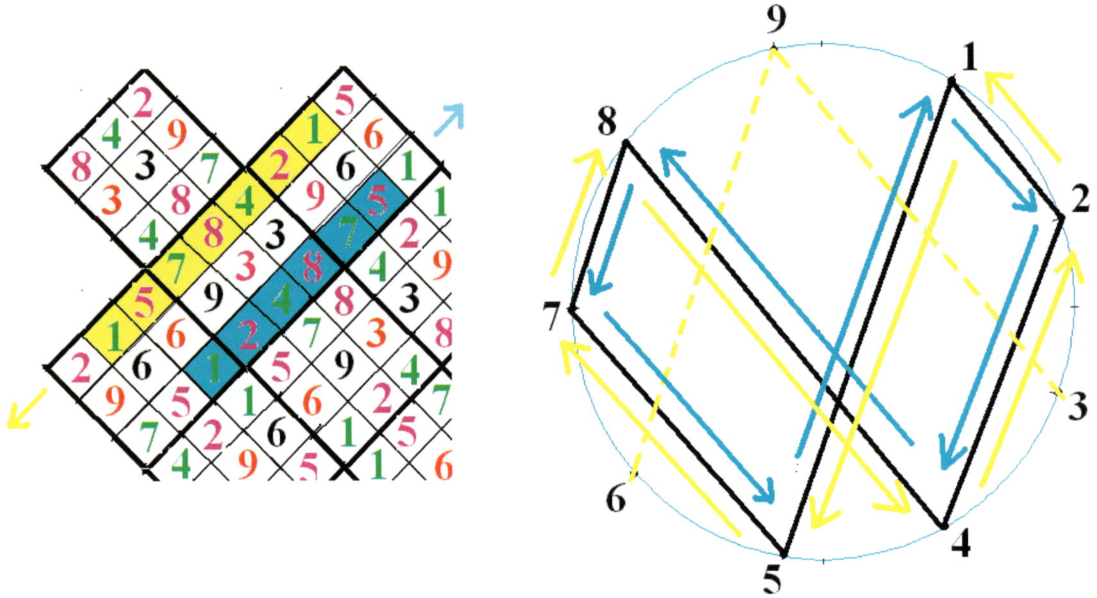

fig. 72 The doubling circuit is represented in the quantum numerology map.

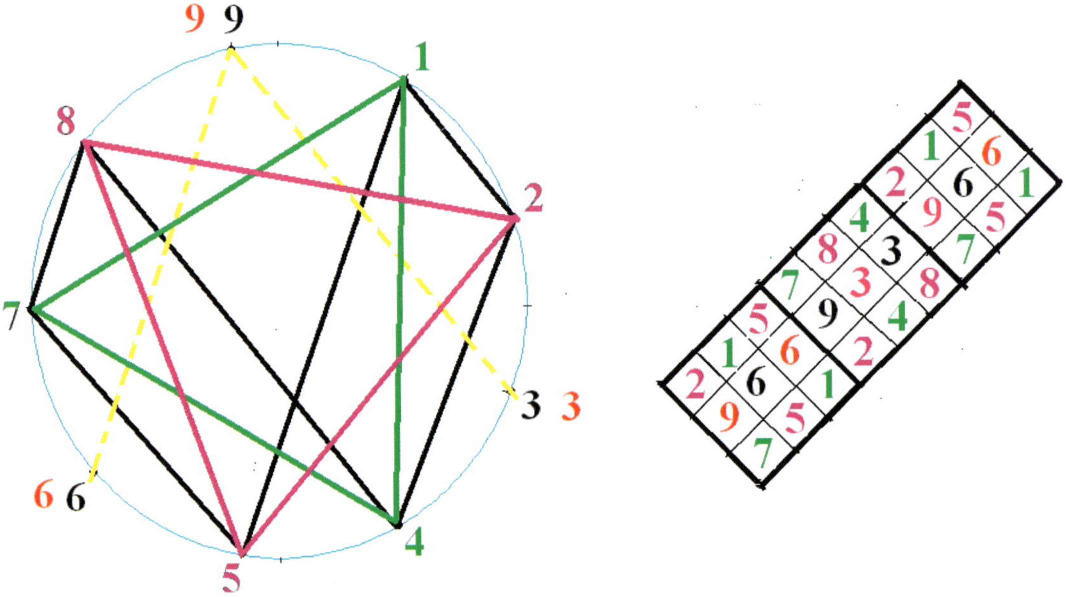

The family numbers: 1 4 7, 2 5 8, 3 9 6, 3 9 6

fig. 72

The polarity of the number 9 determines the polarity of the numbers around. When the number 9 is positive, the numbers 3 and 6 are negative and the numbers: 1,2,4,7,8, and 5 are positive. Vice versa, when the number 9 is negative, the numbers 3 and 6 are positive and the numbers: 1,2,4,7,8, and 5 are negative, see (fig. 73 and 74).

fig. 73 These quadrants, also called vortices, build the quantum numerology map.

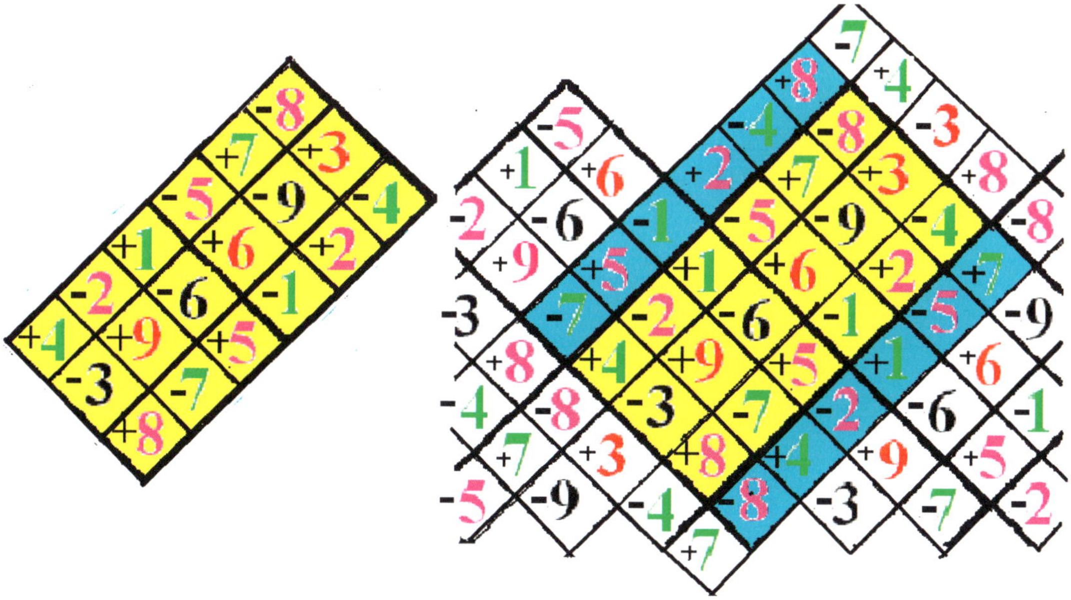

fig. 74

In this map, the multiplication tables are modeled and interconnected on series of numbers. In (fig. 75) we can see the series of polar numbers 1 and 8 starting from the number 9. The series of numbers: 12345,….,9 and the series of numbers: 87654,….,1 in the opposite direction. The polar number 9 gives the image effect to these series of numbers. The numbers 1 and 8 are modeling the vertical axis in Vortex Based Mathematics, see (fig. 75).

fig. 75 The polar number series 1 y 8.

1x8=8
2x8=7
3x8=6
4x8=5
5x8=4
6x8=3
7x8=2
8x8=1
9x9=9

1x1=1
2x1=2
3x1=3
4x1=4
5x1=5
6x1=6
7x1=7
8x1=8
9x1=9

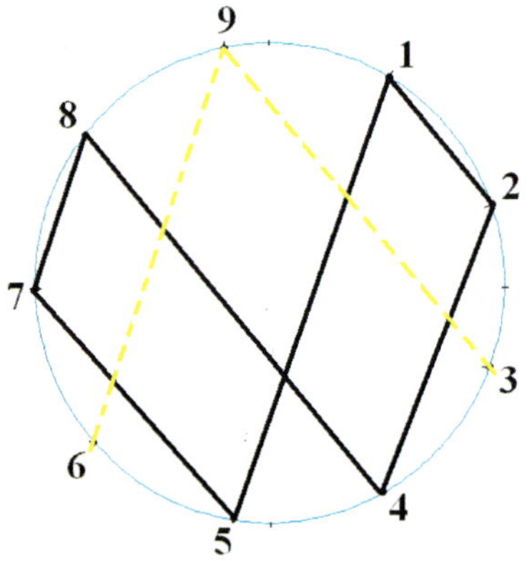

fig. 76 The polar number multiplication tables 1 y 8.

In (fig. 77) we can see the series of polar numbers 5 and 4 starting from the number 9. The series of numbers: 5,1,6,2,7,3,8,4, and 9 and the series of numbers: 4,8,3,7,2,6,1, and 5 in the opposite direction. The number 9 gives the image effect to these series of numbers. The polar numbers 5 and 4 are modeling the horizontal axis in Vortex Based Mathematics, see (fig. 77).

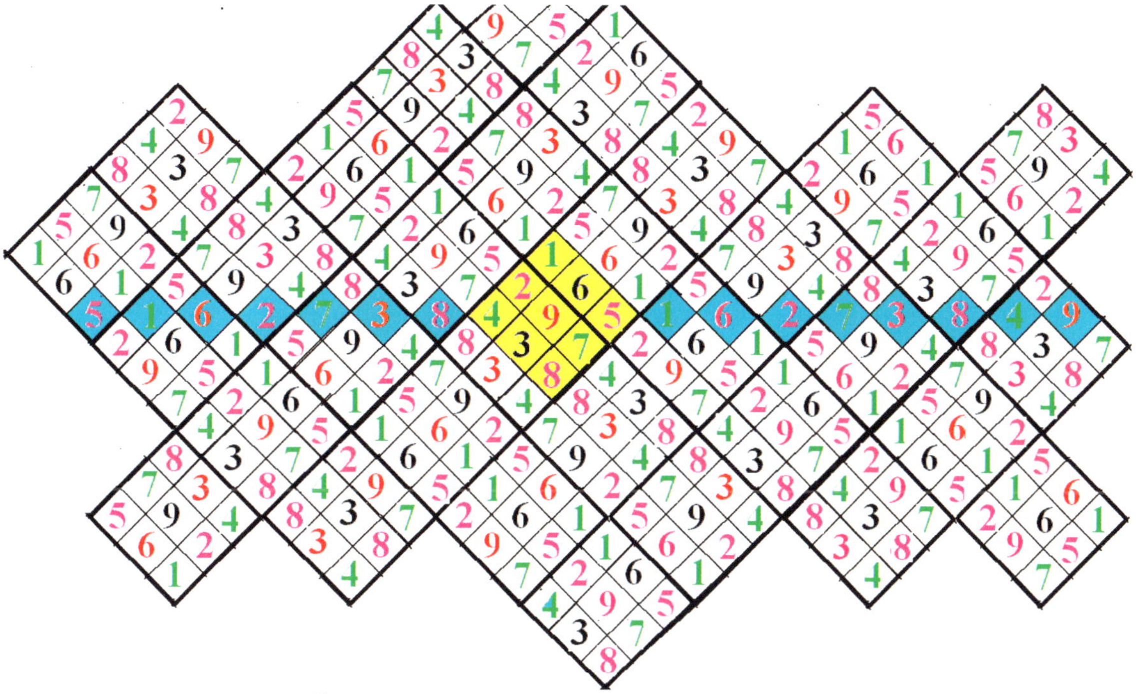

fig. 77 The polar number series 5 y 4.

1x5=5
2x5=1
3x5=6
4x5=2
5x5=7
6x5=3
7x5=8
8x5=4
9x5=9

1x4=4
2x4=8
3x4=3
4x4=7
5x4=2
6x4=6
7x4=1
8x4=5
9x4=9

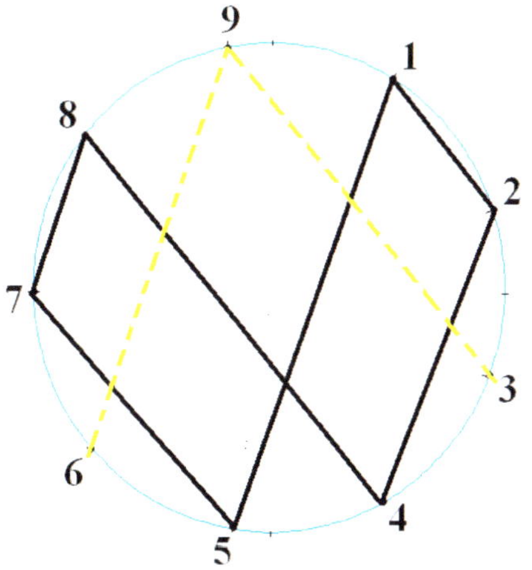

fig. 78 The polar number multiplication tables 5 y 4.

The polar numbers 1 and 8 are modeling the vertical axis, and the polar numbers 5 and 4 are modeling the horizontal axis in Vortex Based Mathematics, see (fig. 79).

fig. 79

In Vortex Based Mathematics everything revolves around the number 9

The polar numbers: 1,2,3,4,5,6,7, and 8 are located around the number 9. As mentioned previously, the polar numbers 1 and 8 are modeling the vertical axis, and the polar numbers 5 and 4 are modeling the horizontal axis. The polar number 2 and 7 represent the z axis*. The z axis is directed from the center of the vortex out, see (fig. 80). The numbers 3 and 6 are the magnetic numbers which create the magnetic field in the vortex, and the number 9 is the emanation* coming from the center of the vortex.

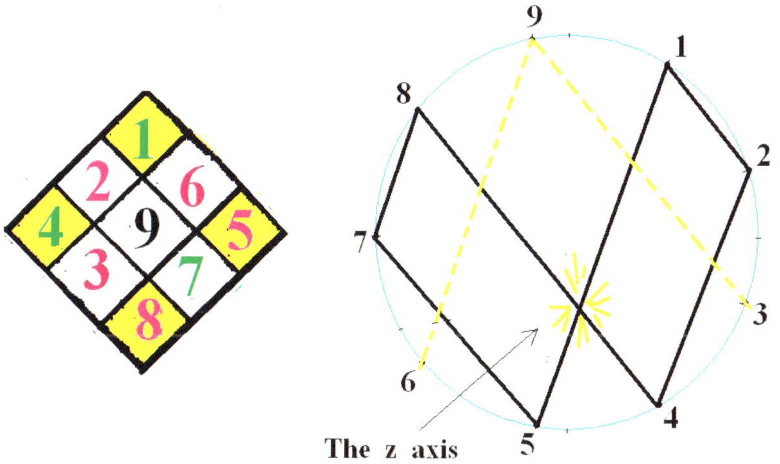

fig. 80 The vortice

When the number 9 changes polarity the numbers around changes its polarity, see (fig. 81).

fig. 81

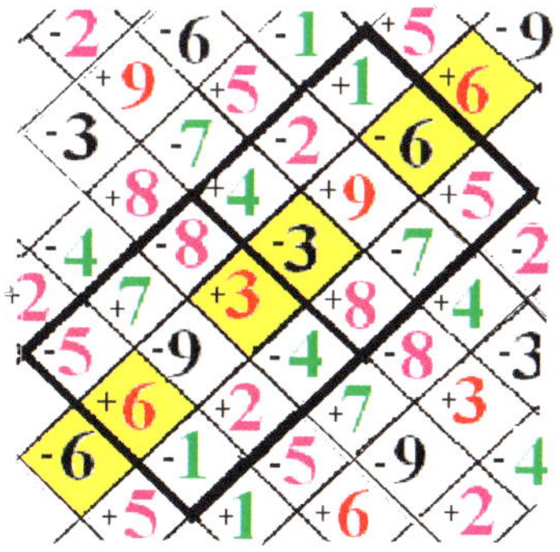

fig. 82

The numbers 6 and 3 repeals each other like magnetic poles, the number 9 is located always between them, see (fig. 82).

*Note: The concepts of Vortex Based Mathematics will be understood as the reader reads more about this subject, this book is presenting only the basic concepts

The (fig.83) shows the doubling circuits flowing simultaneously in opposite directions. These doubling circuits are called doubling circuit images since they are image of one another. We refer to them as circuits because vibration or mass passes through them. The series of numbers: 6 9 3 3 9 6, which are the multiplication tables of the numbers 3, 6, and 9. These series of numbers are found between the doubling circuits, see (fig. 83 and 84).

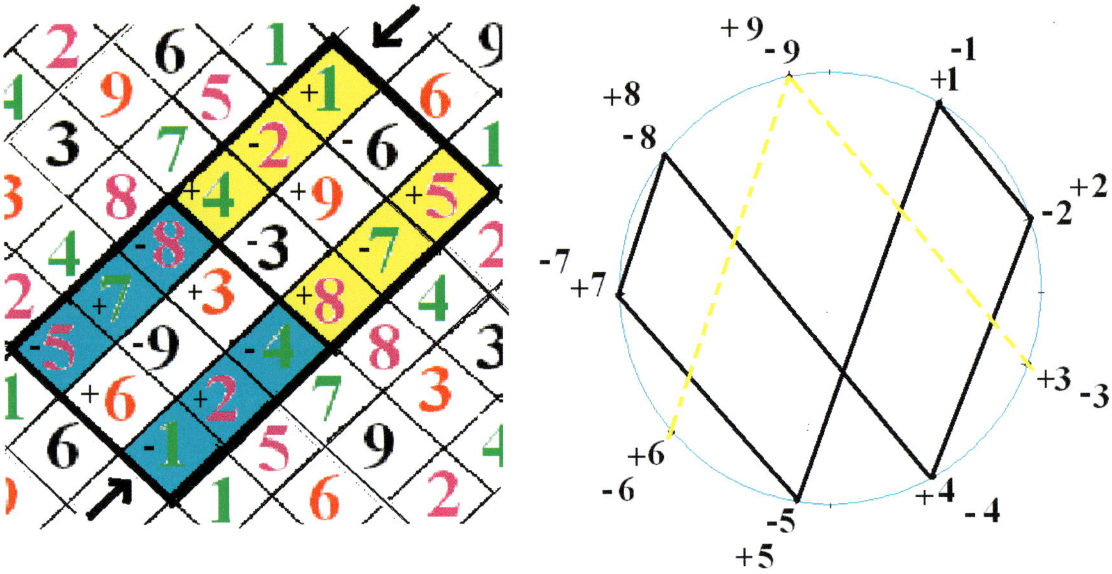

fig. 83

1x3=3	1x6=6	1x9=9
2x3=6	2x6=3	2x9=9
3x3=9	3x6=9	3x9=9
4x3=3	4x6=6	4x9=9
5x3=6	5x6=3	5x9=9
6x3=9	6x6=9	6x9=9
7x3=3	7x6=6	7x9=9
8x3=6	8x6=3	8x9=9
9x3=9	9x6=9	9x9=9

fig. 84

The sum of the numbers that form a perfect square where the number 9 is its center is equal to 9, see (fig. 84, 85, and 86).

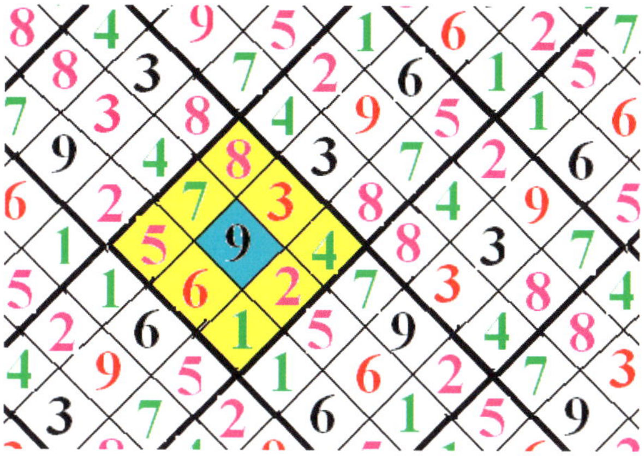

fig. 84 The sum of the numbers around the number 9 is equal to 9
5+7+8+3+4+2+1+6=36~ 3+6=9

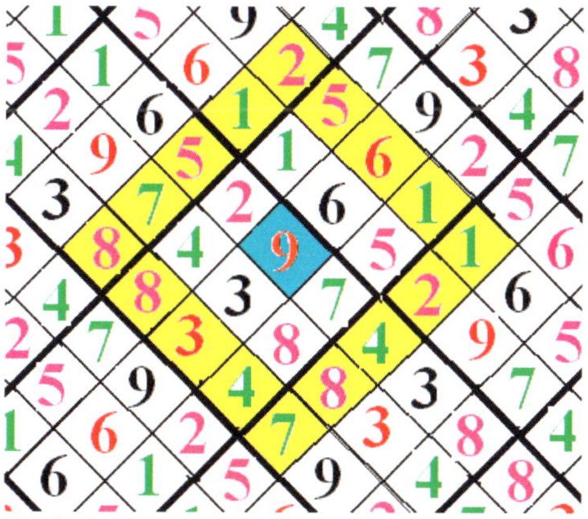

fig. 85 The sum of the numbers around the number 9 is equal to 9
2+5+6+1+1+2+4+8+7+4+3+8+8+7+5+1=72~7+2=9

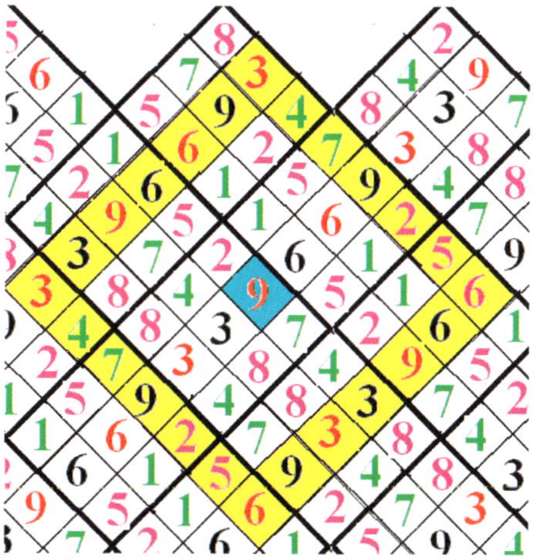

fig. 86 The sum of the numbers around the number 9 is equal to 9
3+9+6+6+9+3+4+7+9+2+5+5+6+6+9+3+3+9+6+5+2+9+7+4+3=9

The family numbers: 1 4 7, 2 5 8, +3+9+6, -3-9-6

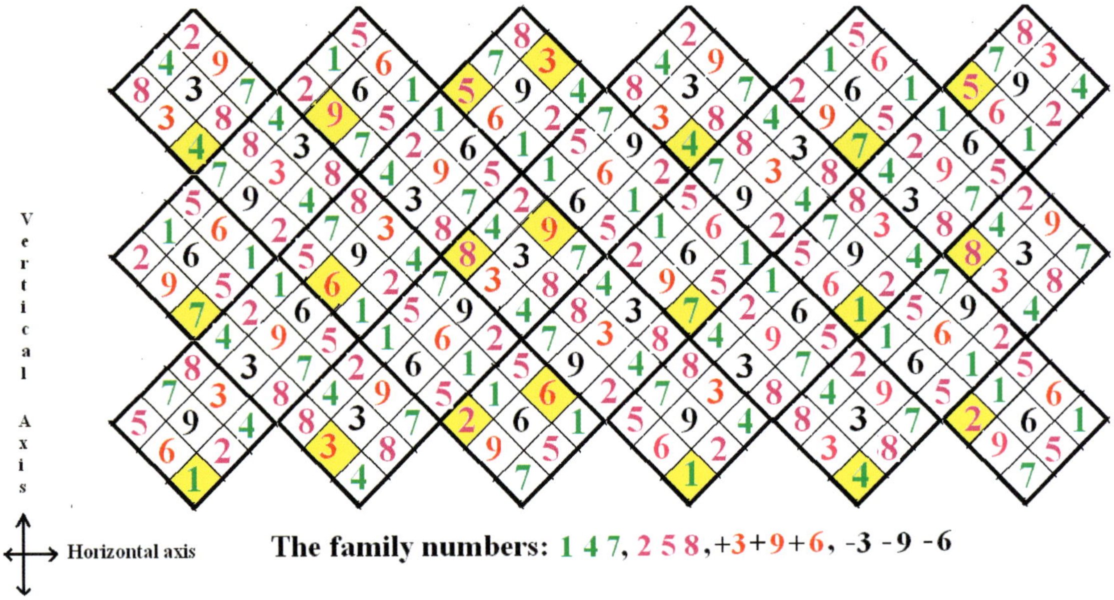

fig. 87 The families of numbers are separated by three spaces on the vertical axis.

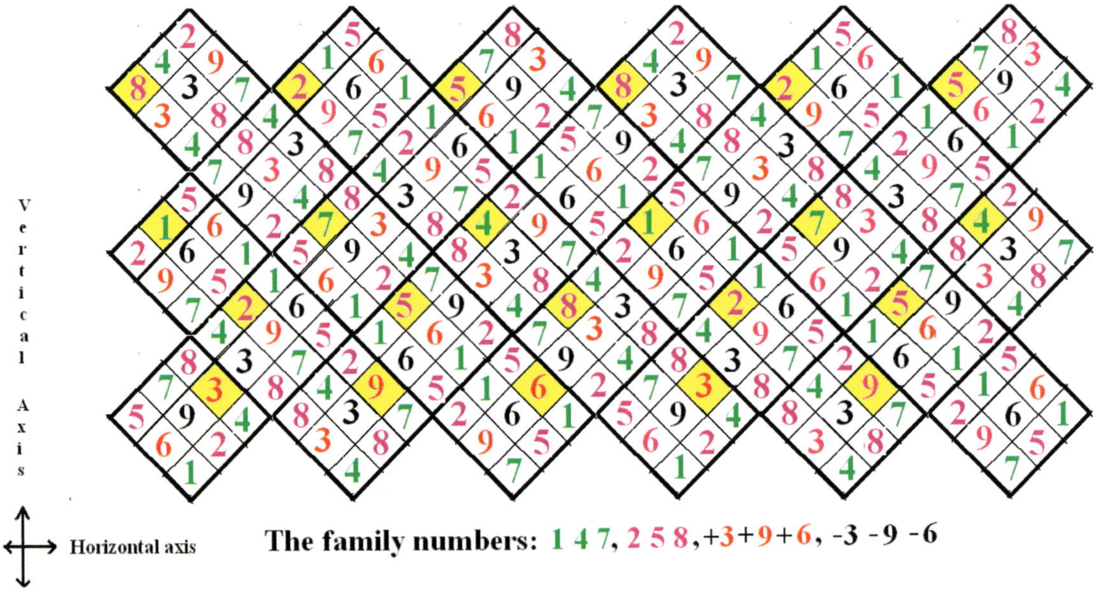

fig. 88 The family numbers are separated by three spaces on the horizontal axis.

The multiplication of the family numbers of neighboring doubling circuits is always equal to one. That is, 2x5 =1, 4x7=28~2+8=10~1+0=1, 8x8=64~1, 7x4=1,5x2=1, 1x1=1, see (fig. 89). This is the property responsible for the harmony* between doubling circuits.

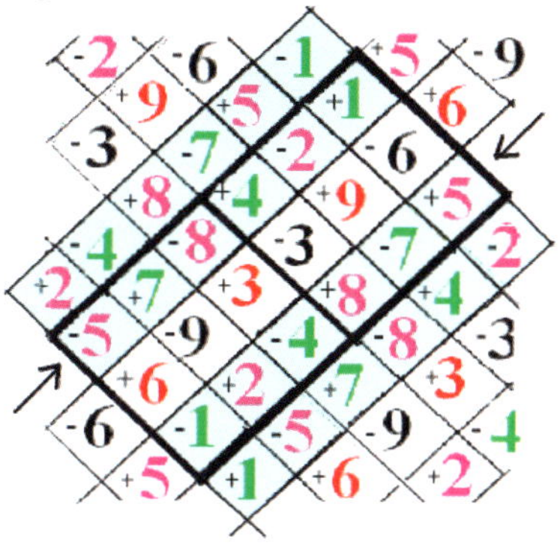

fig. 89

The quantum numerology map is the skin of the vortex which has the shape of a donut; this is the representation of the vortex in the third dimension, see (fig. 90).

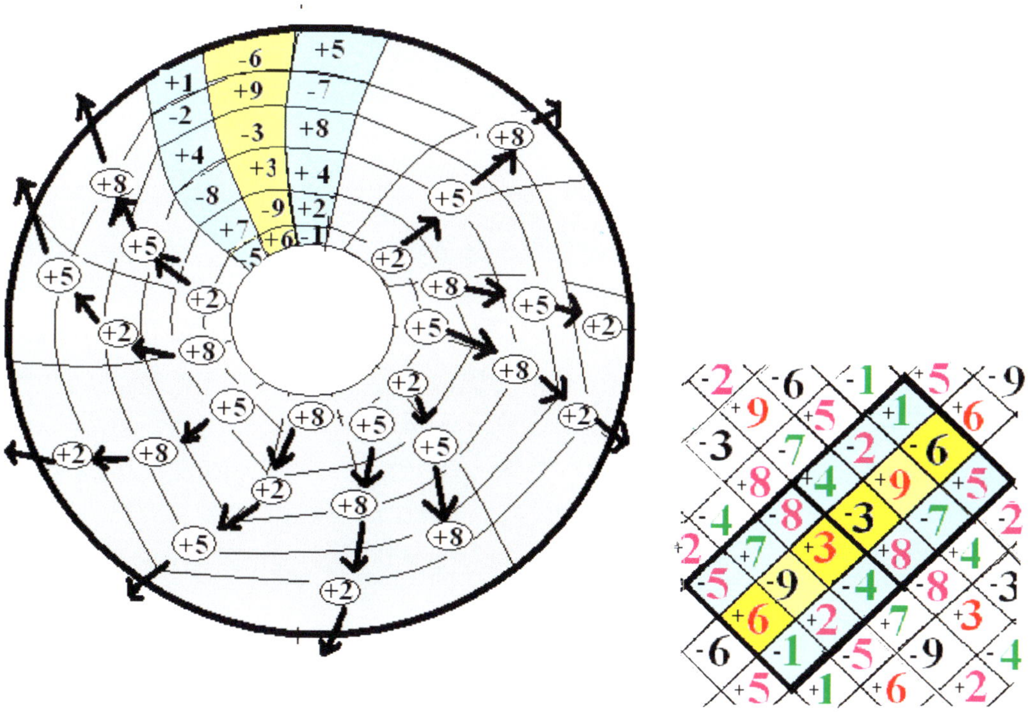

fig. 90

*Note: The concepts of Vortex Based Mathematics will be understood as the reader reads more about this subject, this book is presenting only the basic concepts

Vortex Based Mathematics is a subject of extensive study and its mathematics is complex. Therefore, it is essential to have knowledge of the basic principles. To expand your knowledge of this subject, I recommend reading the extensive material written by Marko Rodin.

Reference

Articles written by Marko Rodin
Lectures on the internet by Marko Rodin and Randy Powell

About the author:
Martha Pieper earned an undergraduate degree in Mechanical and Electrical Engineering at the Institute of Technology in Tijuana, B.C. Mexico (Instituto Tecnologico de Tijuana, B.C. Mexico). At the time of this writing, she is seeking a MS in Mechanical Engineering at the University of Texas at Tyler.

Printed in Dunstable, United Kingdom

91222508R00036

This book is based on the mathematical principles discovered by Marko Rodin and focused only on the basic principles of Vortex Based Mathematics.

ISBN 9780984615353

Survey of Public Library Efforts to Provide Job Information Resources

Primary Research Group

ISBN 978-1-57440-219-3